D1708345

Human Fertility:
The Regulation of Reproduction

Human Fertility:

The Regulation of Reproduction

Celso-Ramón García, M.D.

Division of Human Reproduction
University of Pennsylvania
Department of Obstetrics & Gynecology
Philadelphia, Pennsylvania

David L. Rosenfeld, M.D.

Division of Human Reproduction
University of Pennsylvania
Department of Obstetrics & Gynecology
Philadelphia, Pennsylvania

F. A. DAVIS COMPANY, PHILADELPHIA

Printed in the United States of America.

Library of Congress Cataloging in Publication Data

Garcia, Celso-Ramon, 1921–
Human fertility.

Includes bibliographies and index.
CONTENTS: v. 1. The regulation of reproduction.
1. Birth control. 2. Population policy. 3. Conception—Prevention. 4. Birth control—United States. 5. Population policy—United States. I. Rosenfeld, David L., joint author. II. Title. [DNLM: 1. Family planning. 2. Fertility. 3. Reproduction. HQ766 G216h]
HQ766.G35 301.32'1 76-15411
ISBN 0-8036-3910-4

This book is dedicated
to the workers
in the field of Human Fertility
whose continuing efforts
have yielded an improvement
in the quality of life.

Preface

"It is as much the duty of the physician to inform mankind of the means of preventing evils that are liable to arise from gratifying the reproductive instinct, as it is to inform them how to keep clear of gout or dyspepsia."

(Knowlton, C.: Fruits of Philosophy, Boston, 1833, p. 13)

Although Charles Knowlton, a New England physician, wrote his philosophy in the early nineteenth century, it is only recently that medical schools have begun to teach adequately the responsibility of the physician in aiding his patients in wise family planning.

Problems of reproductive control were felt to be the domain of social scientists, not physicians. Concepts of human sexuality were not considered to be a medical responsibility—the physician was one who cured. He might be involved in disease prevention if he had time or if he was interested, but he did not view social well-being and its ramifications as a part of his health care responsibilities. In fact, he projected from his own life experiences, feelings, moral values, and standards into his clinical situation. (Unfortunately this is true even today in some instances.)

Changes in medical concepts of reproductive control have occurred in recent decades. Early twentieth century activists viewed contraception as a means of increasing the quantity and quality of sexual relationships and thus improving interpersonal relationships through sexual freedom and gratification. The clinical development of the birth control pill by Doctors John Rock, Gregory Pincus and Celso-Ramon García diminished dependency upon permanent contraception and upon cumbersome, less effective, mechanical contraceptives. Today the specialty of family planning is viewed as encompassing the broad aspects of conception and contraception and involving all the implications they impose on the individual, the family and society. Concepts of population dynamics can no longer be ignored.

Contraceptive modalities have been available for centuries but they did not have wide acceptance until they became part of the general health care of the individual. This acceptance resulted particularly because of the high degree of efficacy and follow-up care emphasized by or associated with the distribution of the oral contraceptive and the intrauterine device. Because of a growing recognition that society's ills could be better prevented than treated, it became evident that a cooperative effort must be made by deliverers of health care, whether they be physician, nurse, social worker, family planning worker, educator or teacher. When the health practitioner understands underlying reasons for family planning, he or she can better implement specific approaches.

In recent years the curriculum exposure of medical students to all aspects of medicine has been singularly curtailed. There is greater flexibility and reliance on the students' personal initiative. With this trend has come only brief exposure to problems of reproduction, which are presented in a manner that does not allow the student to develop the broad scope of the problem.

Our text developed from a realization of the need for a single comprehensive family planning education manual. The many textbooks that covered the subject placed greater emphasis on basic reproductive physiological principles. A single, specific, encompassing, yet easily readable, educational source was not available. The student had to rely on journal articles or specific chapters in lengthy books. The result was a perpetuation of much of the student's prior information. These adverse features should be reduced with this book since the average total reading time should not exceed more than ten hours. Graphic information, which is sprinkled liberally throughout the text, gives additional data that do not require lengthy concentrated reading and allow retention of information for longer periods of time.

It should be emphasized that this book is specifically intended to serve as an introductory textbook on which the student may develop his ideas. It can serve as an easy reference for the established practitioner. Moreover it should be of value not only to the deliverer of health care but also to selected college and high school students as a source of material on population dynamics and means of better understanding of sexuality. Concepts are purposely presented in outline form in conjunction with graphs, tables and illustrations. The ideas in the text are meant to be questioned, reasoned out, developed and pursued either within the classroom discussion or by the reader through the use of specific references and bibliographies.

The first three chapters deal specifically with population dynamics. Basic demographic concepts are introduced and the problems of excess reproduction at both an international and national level are explored. The importance of family planning in the fulfillment of social needs is discussed. It should be remembered that individuals form families and families form society. It is the integration of family units and societal groups with their interplay which is responsible for the population dynamics and, therefore, the welfare of the world. Thus, in Chapters 4, 5 and 6 we deal with the specific needs of the individual who can be helped through available family planning services. Pressures on and within the family unit, the sexually active adolescent and the relationship between available contraception and the ability of women to find complete social liberalization are reviewed.

Chapters 7 and 8 emphasize the medical approaches to fertility control. The significance of contraception is portrayed. Presently available contraceptive modalities are discussed relative to their particular advantages and disadvantages and specific clinical situations are described. Finally, surgical termination of pregnancy, which is primarily viewed as a back-up to failed contraception, is discussed within a historical and clinical perspective.

The principal title of this textbook, *Human Fertility,* has a subtitle *The Regulation of Reproduction.* A series on Human Fertility will continue with similarly presented books encompassing the management of the infertile couple, the reproductive physiology and endocrinology of reproduction and the reproductive considerations of sociologic and emotional implications in interpersonal relationships. This series will represent the efforts in the main of members of the Faculty of Medicine at the University of Pennsylvania.

Acknowledgments

It is difficult to express our appreciation to all the persons who had a significant role in the preparation of this book. We are indeed indebted for the thoughts and suggestions of many, especially Dr. John Rock, Professor Emeritus in Gynecology at Harvard Medical School, Dr. Luigi Mastroianni, Professor and Chairman of the Department of Obstetrics and Gynecology at the University of Pennsylvania, Dr. Elaine Pierson, Student Health Officer and member of the Department of Obstetrics and Gynecology at the University of Pennsylvania and Mrs. Ann Jane Levinson, Dean, Columbia School for Girls, New York City.

The people who have given us permission to use their material liberally and who have given us advice along the way are so numerous that it would be difficult to list them all. In particular we would like to mention Ms. Phyllis Pietro of the George Washington University Population Information Bureau, Ms. Lynn Landman and Mr. Dick Lincoln of the Alan Gutmacher Institute and Coeditors of the Family Planning Perspectives, Mr. Robin Elliot and Dr. Louise Tyrer of Planned Parenthood, Dr. Elizabeth Connell of the Rockefeller Foundation and Mr. Christopher Tietze of the Population Council. Our special thanks to the Holland Rantos Company for permission to use the illustrations concerning insertion of the diaphragm.

The illustrators, Ms. Sandy Mayer, Ms. Laurie Mitchell and Ms. Pat Horoshak deserve special commendations as well as our two secretaries, Mrs. Dorothy (Sue) Cooper and Ms. Paula Branca, who worked so diligently to put this volume together.

To our publishers we are indeed indebted for the able editorial efforts of Mrs. Christine Young and Mrs. Eleanor Mora and for the continuing emotional support given us by Mr. Robert H. Craven, President.

Last but not least and probably the most critical and helpful have been the numerous medical students who have lived through our elective in Human Reproduction, whose interest, enthusiasm and curiosity led to the formation of this book and whose criticisms have aided greatly in the finished product.

For the patience shown by our mates, we are thankful; we are the better for our experiences.

Celso-Ramón García, M.D.
David L. Rosenfeld, M.D.

Contents

1	Demographic Factors in Fertility	1
2	Population Problems in the World	7
3	Population Problems in the United States	23
4	Population Stresses on the Family Unit	31
5	The Changing Status of Women	39
6	Adolescent Sexuality	47
7	Contraceptive Methods	59
8	Pregnancy Termination	123
Index of Subject Matter and Illustrations		155

1 Demographic Factors in Fertility

Changes in fertility of a specific population are influenced by a number of interrelated factors. Included in these variables are health care, social pressures, prevailing economic conditions, religious beliefs, historical and political trends, cultural patterns and education.

The systematic variations in fertility which are observed within or between populations is known as *differential fertility*. There would be no differential fertility if every woman gave birth to an identical number of children. There is much woman-to-woman variation, however, and it is the factors behind these variations that demographers study.

The demographer categorizes the population into demographic, social or economic groupings. Fertility is then computed for each category. These studies of differential fertility can serve to identify high fertility segments of the population and can direct family planning workers to these groups. The studies can also be a predictive function for workers in the field of population dynamics, sociology and anthropology.

SELECTED CATEGORIES

Several categories in demographic factors of fertility are presented here in outline form with selected samples used for demonstrative purposes. In reviewing each example, the student should attempt to analyze underlying reasons for the facts.

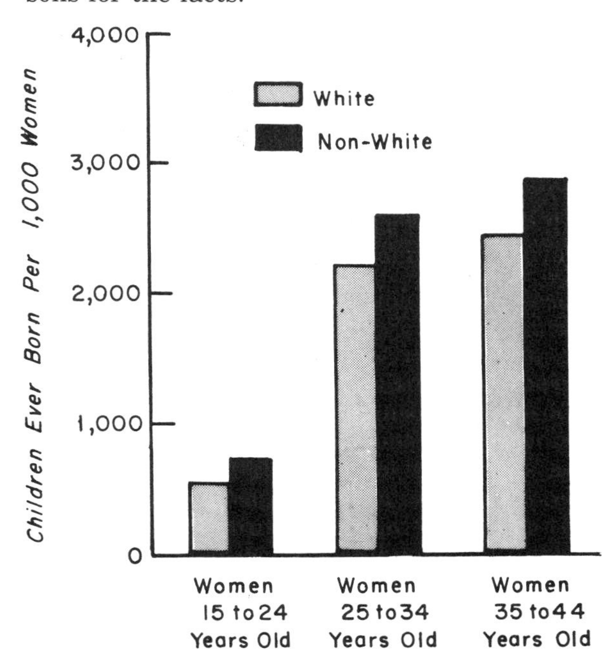

FIGURE 1-1. Number of children ever born per 1000 women aged 15 to 44 years, in the U. S., by age and race, 1960. (Adapted from Smith, T. L., and Zopf, P. E.: *Demography: Principles and Methods*. F. A. Davis Co., Philadelphia, 1970, p. 337.)

I. Age—the most important demographic correlate of fertility (Fig. 1-1). (See Chapters 5 and 6)

II. Ethnicity
   A. American Indians are the most fertile of non-white racial groups.
   B. With the exception of Mexicans, foreign-born white females have lower fertility than native-born white female Americans.

III. Residence
   A. In the United States, the Northeast has the least fertile population.
   B. Black females living in the South have the highest fertility.
   C. Rural females have higher fertility than suburban females, who in turn are more fertile than urban females.

IV. Education
   A. Fertility rates are inversely related to education, especially with blacks.
   B. Those countries with high illiteracy rates tend to be those with higher fertility rates. These countries also show marked discrepancy between male and female literacy (Fig. 1-2).

V. Marital Patterns
   A. Disruption in marriage depresses fertility.
   B. Married white couples tend to concentrate their childbearing within the first 10 years of married life.
   C. The younger the husband is in comparison with the wife, the greater the tendency toward larger families.

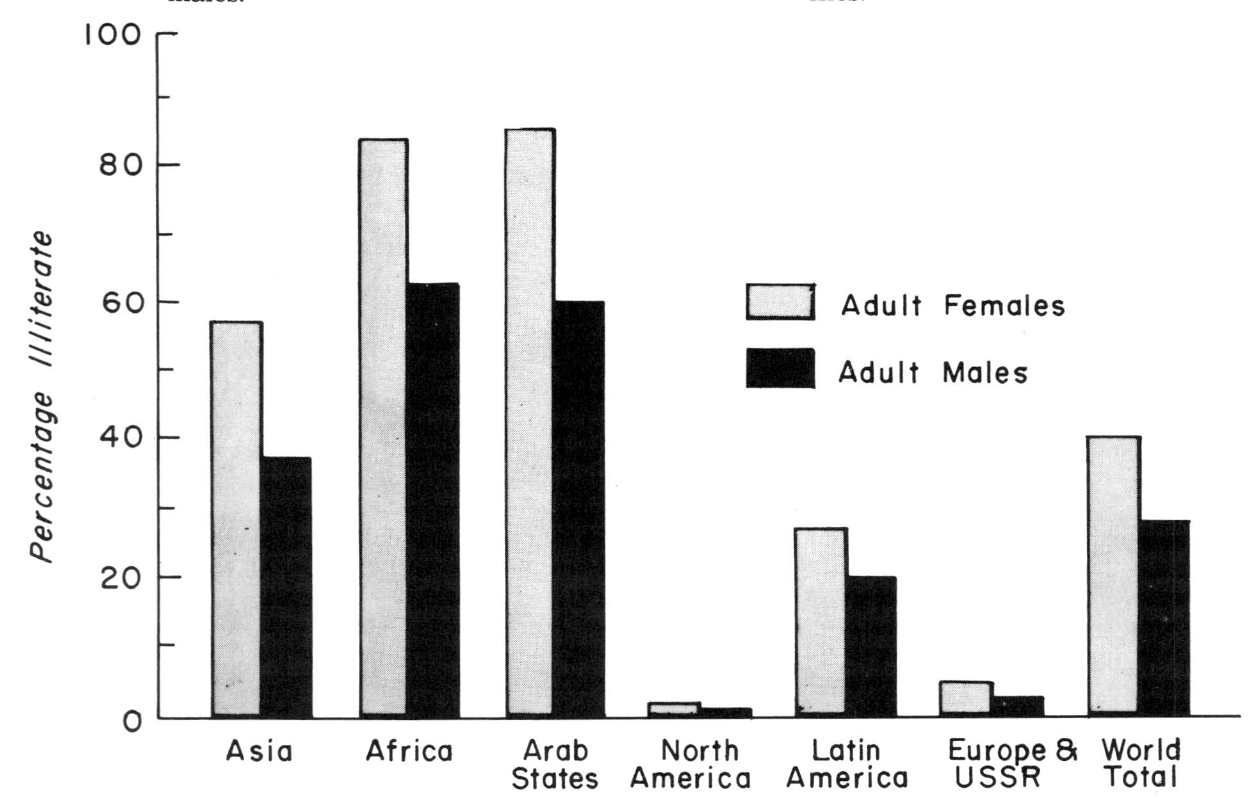

FIGURE 1-2. Male and female adult illiteracy rates from seven countries, around 1970. (Adapted from Population Bulletin, Population Reference Bureau, Vol. 30, No. 2, 1975, p. 9.)

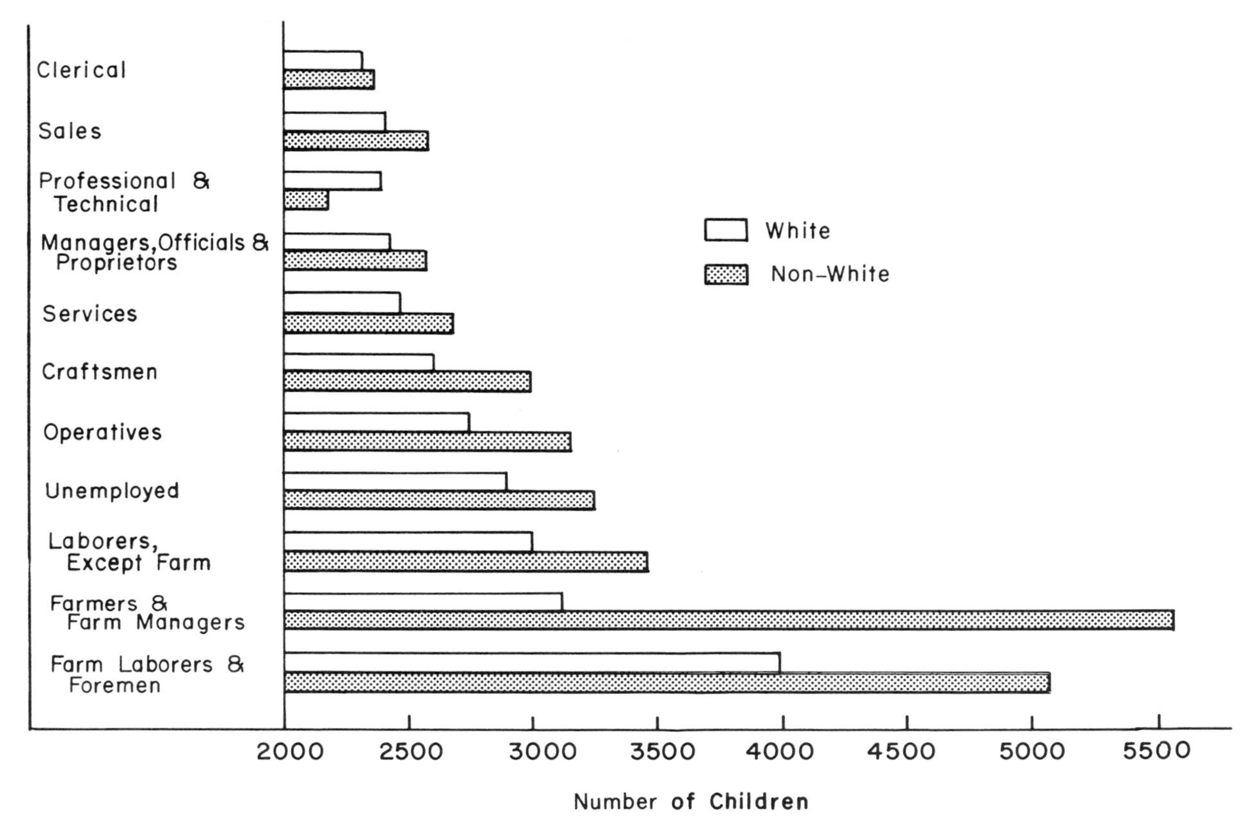

FIGURE 1-3. Number of U. S. children ever born per 1000 married women aged 35 to 44, by occupation of husband and race, 1960. (Adapted from Smith, T. L. and Zopf, P. E.: *Demography: Principles and Methods.* F. A. Davis Co., Philadelphia, 1970, p. 349.)

VI. Migration—migrants tend to be less fertile than non-migrants.

VII. Occupation—there is a weak inverse relationship between occupational status and fertility (Fig. 1-3).

VIII. Income—higher income for whites has resulted in a direct relationship with their fertility. This contributed in part to the baby boom of the 50's. This pattern is inverse for blacks.

In reviewing the above it should be appreciated that (1) *these relate to general trends and are not specific for the individual,* and (2) *there is a marked interrelationship among all these variables in their influence on fertility patterns.*

ELEMENTS OF DEMOGRAPHY

The three basic elements in the science of demography include *basic theoretical issues, collection and evaluation of data* and *basic methodological tools.* The following sections demonstrate examples of these elements as applied to population growth.

Basic Theoretical Issues

Examples of these theories are the concept of population as an ecosystem variable and the Malthusian, Marxist and Kingsley Davis theories.

Population as an Ecosystem Variable. This concept assumes the effects of population balance and population optima on the environment and on political, social and economic institutions.

Malthusian Theory. This theory was considered in 1798 in an essay on population by an English economist. It postulates that there is a universal tendency for population to outrun its means of subsistence. It further states that human reproduction progresses geometrically, thereby outrunning food resources which progress arithmetically. The criticism of this theory is that it does not account man's ability to accommodate, his capacity to open new horizons and his capacity to expand his resources.

Marxist Theory. Marxism, as developed by Marx and Engels, gives class struggle a primary role in eliminating capitalism in favor of socialism. Concerning population growth, it theorizes that (1) population growth is not subject to any immutable natural law; (2) population growth is a result of social and economic conditions; and (3) surplus population is relative to the distribution of capital, an inevitable product of the capitalistic system.

The Marxist theory was one of the basic points of discussion and disagreement which enveloped the recent Bucharest World Population Conference.

Kingsley Davis Theory. A theory of multiphasic response that states there is a complex nature of demographic variation. It further reasons that reproductive behavior is influenced to a great extent by social motivation. He contends

> If it were admitted that the creation and care of new human beings is socially motivated, like other forms of behavior, by being a part of the system of rewards and punishments that is built into human relationships, and thus is bound up with the individual's economic and personal interests, it would be apparent that the social structure and economy must be changed before a deliberate reduction in the birth rate can be achieved.[1]

In considering whether population control should be managed by medical personnel or sociologists, he says

> Designation of population control as a medical or public health task leads to a similar evasion. This categorization assures popular support because it puts population policy in the hands of respected medical personnel, but, by the same token, it gives responsibility for leadership to people who think in terms of clinics and patients, of pills and IUD's, and who bring to the handling of economic and social phenomena a self-confident naivete. The study of social organization is a technical field; an action program based on intuition is no more apt to succeed in the control of human beings than it is in the area of bacterial or viral control.[2]

Kingsley Davis' theory emphasizes the need, as shown in this book, for a total involvement of all social scientists in the field of population.

Collection and Evaluation of Data

Data collection is the foundation of demographic study. One must be able to understand the basic tools with which the demographer works. These are his sources of data. They include the census, vital registration and surveys. However, his data are not infallible. There are defects and errors in any form of data collecting. One must also be able to understand the limitations inherent in the means of data collection and in the data itself.

Basic Methodological Tools

Any technical field has a working vocabulary. The more commonly used demographic terms are given here as a necessary tool for anyone studying demographic factors in fertility.

crude birth rate—yearly number of live births per 1000 population

crude death rate—yearly number of deaths per 1000 population

crude rate of natural increase—difference in the yearly number of births and deaths per 1000 population

rate of population increase—the yearly change in population size due to the rate of natural increase plus migration, expressed as a percentage of the total population

general fertility rate—the number of live births per 1000 females aged 15 to 44

age specific fertility rate—the yearly number of live births in a specific age grouping per 1000 females in that age

gross reproduction rate—the sum of all age specific birth rates for a given year times the proportion of all births that are female, that is, the average number of daughters born to a woman

total fertility rate—the sum of the age specific birth rates over the entire childbearing years, that is, the average number of live children born to a woman

net reproduction rate—similar to gross reproduction rate but reduced to reflect the failure of all females to survive their childbearing years, indicating how one generation of childbearers is being replaced by the next one. When NRR = 1.0, replacement fertility is achieved.

dependency ratio—the number of people aged younger than 15 and older than 64 per 100 people aged 15 to 64 in a given population

zero population growth (ZPG)—condition in which, in a given period of time, the population neither increases nor decreases

years to double population—given a fixed crude rate of natural increase, the number of years it would take to double a given population. (Note the inverse relationship in Figure 1-4).

In studying these methodological tools, there are important differences noted in *period fertility* (yearly), which is affected by changing conditions such as later age at childbearing, and *cohort fertility* (the number of children born throughout the entire reproductive span).

DEMOGRAPHIC TRANSITION

A final concept which should be understood for the analysis of historical and current trends in the world population is the concept of the demographic transition. Basically, there are four stages (Fig. 1-5):

1. high and almost equal birth and death rates
2. high rate of population growth resulting from a drop in mortality with a lagging decline in fertility

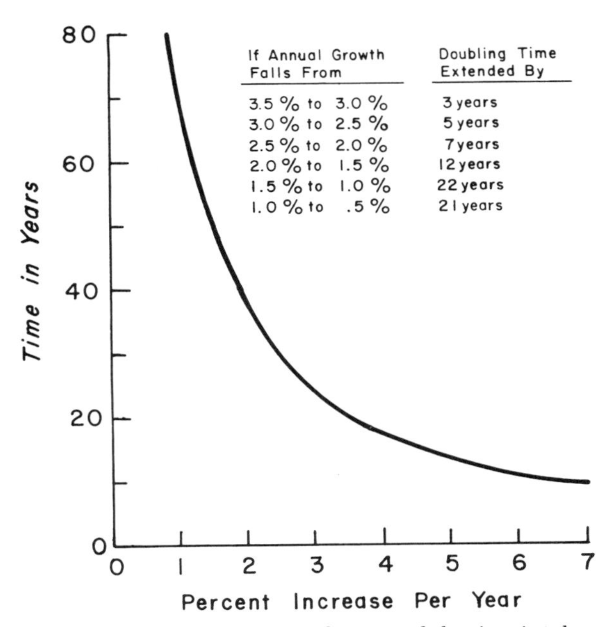

FIGURE 1-4. Population growth rate and the time it takes population to double. Notice in the column on the right how quickly doubling time is extended although in the left column the growth rate falls uniformly each time by 0.5 percent. (Adapted from Nortman, D.: Reports on Population/Family Planning 2:4, 1973.)

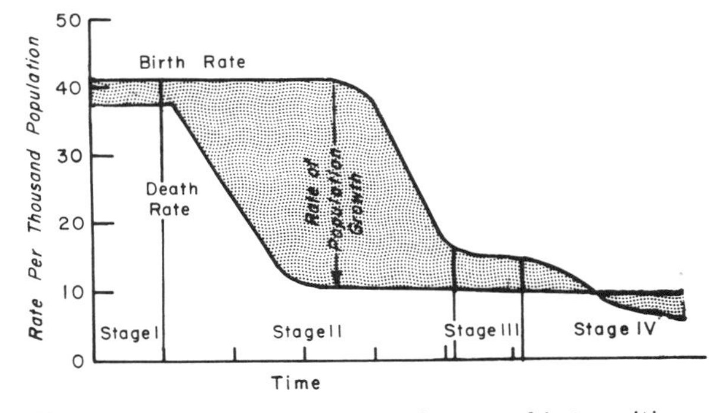

FIGURE 1-5. The four stages in demographic transition.

3. low birth and death rates with a declining rate of growth
4. birth rates which are lower than death rates yielding a population growth deficit

The first stage reflects a situation which existed throughout most of the world's history until the present. The second stage represents the current conditions in most developing countries, while the third stage reflects the current conditions in most developed countries. The fourth stage, which by definition must be temporary, may be necessary to create a stabilization of the population in reaching an adequate level of stage three, and is ideally represented by a zero population growth. This stage was reached by the Federal Republic of Germany in 1974.

With a basic understanding of demography applied to population change, various factors influencing these changes and the means of studying these changes, we shall proceed in further chapters to review population problems at the international, national and family levels.

SUMMARY

Provision of contraception alone is not the answer in dealing with a country's fertility problems. Changes in fertility of a specific population are the result of a number of interrelated factors. An understanding of these factors is essential if one is to attempt to provide family planning services to that population. Demography aids in establishing trends in population growth or decline and studies the consequences of these trends.

QUESTIONS

1. What is needed for a country to reach zero population growth?
2. What factors would be most relevant for the determination of fertility patterns in a developed country? in a developing country?
3. What are the major criticisms of the Malthusian theory? the Marxist theory?
4. What are the factors responsible for the rapid population growth in many developing countries?

REFERENCES

1. Kingsley Davis: *Population policy: will current progress succeed?* Science 158:731, 1967.
2. Ibid.

BIBLIOGRAPHY

Cho, L. J., Grobill, W. H., and Bogue, D. J.: *Differential Current Fertility in the United States.* Community and Family Study Center, University of Chicago, 1970.

Smith, T. L., and Zopf, P. E.: *Demography: Principles and Methods.* F. A. Davis Co., Philadelphia, 1970.

2 Population Problems in the World

Currently there is concern over the rapid growth of the world population and the problems inherent therein.

UN Declaration of Human Rights (1962):

> We believe that the population problem must be recognized as a principal element in long range national planning if governments are to achieve their economic goals and fulfill the aspirations of their people.
>
> We believe that the great majority of parents desire to have the knowledge and means to plan their families; that the opportunity to decide the number and spacing of children is a basic human right.
>
> We believe that lasting and meaningful peace will depend to a considerable measure upon how the challenge of population growth is met.
>
> We believe that the objective of family planning is the enrichment of human life, not its restriction; that family planning, by assuring greater opportunity to each person, frees man to attain his individual dignity and reach his full potential.

Philip Hauser in *The Population Dilemma*[1]:

> Given the present outlook, only the faithful who believe in miracles from heaven, the optimistic who anticipate super-wonders from science, the parochials who expect they can continue to exist in islands of affluence in a sea of world poverty, and the naive who can anticipate nothing can look to the end of this century with equanimity.

Concern over the "population explosion" is reflected in the World Population Conference held in Bucharest in August, 1974.

Final Statement from Bucharest[2]

The following "Statement from Bucharest" was drafted by a group of nongovernment organizations attending the Tribune. Though unofficial, it gathered wide support.

The signatories of this declaration having attended the World Population Conference and Population Tribune, August, 1974, in Bucharest, Romania, believe the following principles should guide the formulation and implementation of policies to achieve a beneficial balance between the world's population and the resources of the earth.

1. The world situation is potentially disastrous. Hundreds of millions are suffering from hunger, poverty, persecution, disease and illiteracy. The unprecedented rate of population growth, doubling population from 3 to 6 billion in a generation, will strain the environment and man's social, political, and economic institutions to breaking point. Action to meet this challenge is imperative.

2. The basis for tackling these problems must be in terms of biological systems and environmental imperatives, justice, equality and the recognition of the dignity of the individual. Radical changes in the world's social, political and economic structure, long overdue on moral grounds, have now become necessary on ecological grounds.

3. In the distribution and consumption of the world's resources, the needs of all social and ethnic groups must be considered and we reject as destructive all policies that are purely nationalistic.

Each nation should take the responsibility of meeting its particular population problems in the interest of the health and well-being of its own people, but having regard to the needs of the people of other countries.

Some countries consume and waste the earth's resources at a rate that cannot be maintained. Others have densely settled regions with population growth rates of 2 or 3 per cent a year that will exert demands on the international community which may not be met. The urgency of the global crisis must not be ignored nor submerged beneath national ambitions.

4. All countries should practice rigorous conservation measures to prevent pollution and waste of both non-renewable and renewable resources, especially food. Among other things, there must be curtailment in the consumption of luxury and expendable items and a greater emphasis on life styles which stress social rather than material values.

5. Population policies should aim to enhance the quality of life of all people. These policies should be integrated into the framework of overall economic and social development policies designed to attack social and economic injustice between and within countries as the world moves toward a new economic order. In population policies the needs of all social and ethnic groups must be considered with special reference to the rights of indigenous and minority groups to take part in such policy formation.

Population problems cannot be solved in isolation from social and economic development. On the other hand, economic development cannot be relied upon by itself to prevent those population problems.

6. Governments must respect the basic rights of all individuals to have access to information and means of determining the number and spacing of their children. By the year 1985, governments should provide free information and services to ensure this right and adequate education on population dynamics to all.

7. We call upon the highly industrialized nations, and other nations that have the capacity, to work with developing countries towards terms of international trade that are just and realistic, and to give suitable aid without strings attached. Food and fertilizers should not be used as economic and political instruments in world trade and international relations.

8. The success of population policies depends on the full participation of women. Women should have the opportunities to obtain full human dignity necessary for them to exercise responsible choice as persons. Governments should take particular steps to achieve integration into every stage of the development process of the second development decade. The economic contribution of women as mothers and providers of food both in rural and urban areas should be fully recognized.

9. Every child born should be assured the conditions within which its full potentialities can be realized. Today millions of children suffer from malnutrition, deprivation and disease. Without assurance that their children will survive, parents lack the first condition of hope and security to freely participate in family planning. Agricultural and health resources must be developed to assure adequate pre- and post-natal health for all children.

10. The non-productive commitment of wealth, knowledge and skill to large military programmes is an offence against humanity and an obstinate barrier to meeting the human needs set forth in the World Population Plan of Action. We call on citizens to work in their respective countries for a reallocation of public funds from the means of destruction to the support of life.

11. We fear that the World Plan of Action will remain impotent in the absence of the necessary changes in the political structures. Large-scale political and social changes will be forced upon mankind. Rather than walking blindly, we must prepare for them responsibly. We urge our fellow participants at these meetings to insist upon realistic discussion of population problems in political terms, and to reject, actively oppose, and go beyond official declarations on population matters whose generality makes them vacuous.

12. We accept personal and organizational responsibility to press these values and decisions on governments and agencies (both governmental and non-governmental), in whatever ways are open to them and jointly and severally to cooperate with the UN Fund for Population Activities and other relevant UN agencies. We believe that striving for these goals will ensure a better life for generations to come.

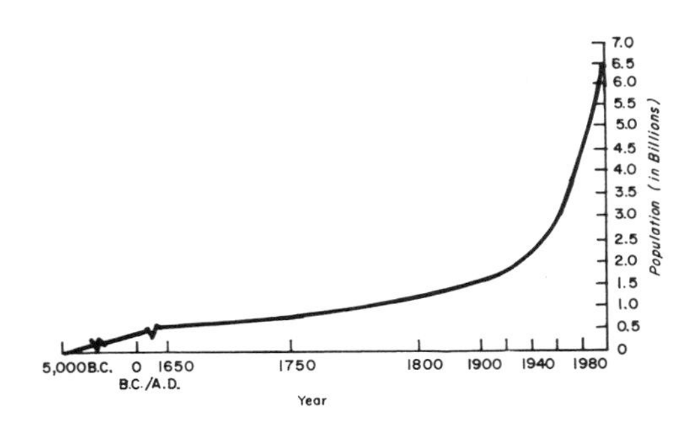

FIGURE 2-1. Growth of world population throughout history. (Adapted from Berelson, B.: *World population: status report 1974*. Reports on Population/Family Planning. January 1974, p. 3.)

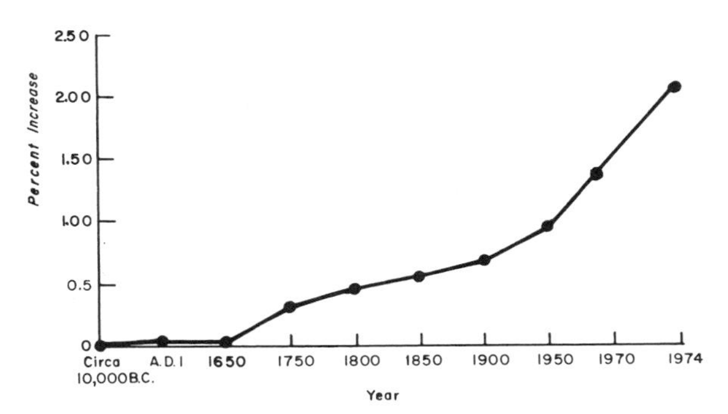

FIGURE 2-2. Estimated annual percentage increase of population. (Based on data from Berelson, B.: *World population: status report 1974*. Reports on Population/Family Planning. January 1974, p. 15.)

At the present time there are 3.9 billion people in the world with a world growth rate of 2 percent yearly or 75 to 80 million people per year with a doubling time of 35 years. The present population growth rate is the highest in history; note the drastic increase (Fig. 2-1). Figure 2-2 shows the percentage increase of population. Again note the sharp increase in recent centuries. Of interest to the demographer is the doubling time for population growth so clearly shown in Figure 2-3.

CONCERNS FOR THE FUTURE

Predictions of future world population size are made on the basis of various assumptions. Study Figure 2-4 to see the important change

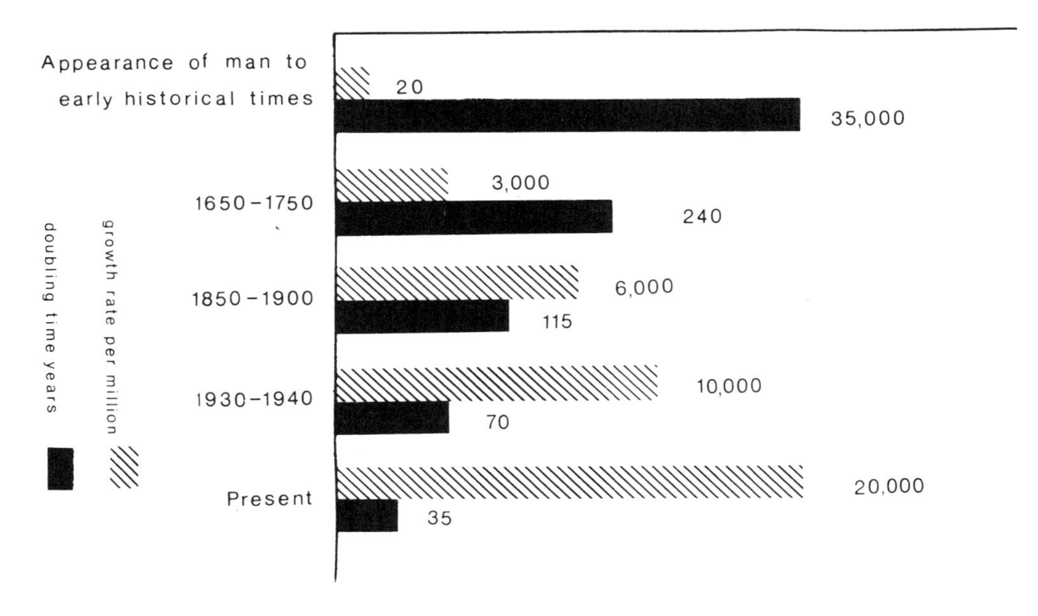

FIGURE 2-3. Population growth rate and doubling time. (Adapted from Berelson, B.: *World population: status report 1974*. Reports on Population/Family Planning. January 1974, p. 15.)

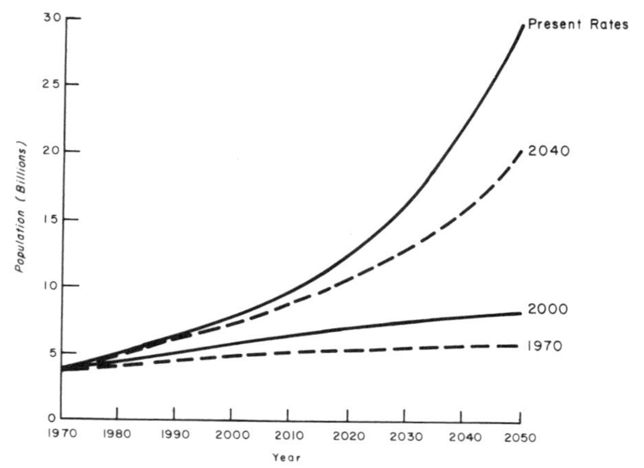

FIGURE 2-4. Projections of world population based on assumptions that replacement fertility (net reproduction rate of 1) was achieved by 1970, by 2000, by 2040 or at present growth rates. (Adapted from Frejka, T.: *The prospects for a stationary world population.* Sci. Am. 228:15, 1973.)

that can occur during the next century at different rates of growth. On the following pages we have shown various factors affecting growth.

Mortality

Such unprecedented growth, as shown in Figure 2-4, is a result of a marked decrease in mortality without a similar decrease in fertility. In the past 30 years the world's crude death rate has halved. For example, in Peru the crude death rate in 1876 was 33 and the crude birth rate was 44, but in 1961 the crude death rate was 15 while the crude birth rate was 45. The very high growth rate in Latin America reflects a death rate lower than most underdeveloped countries but it is accompanied by a high birth rate.

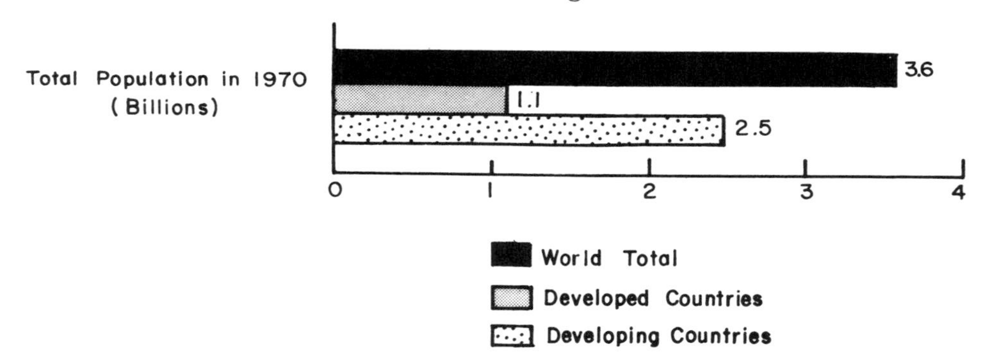

FIGURE 2-5. Total population 1970 in billions for world and for developed and developing countries. (Based on data from Frejka, T.: *The prospects for a stationary world population.* Sci. Am. 228:23, 1973.)

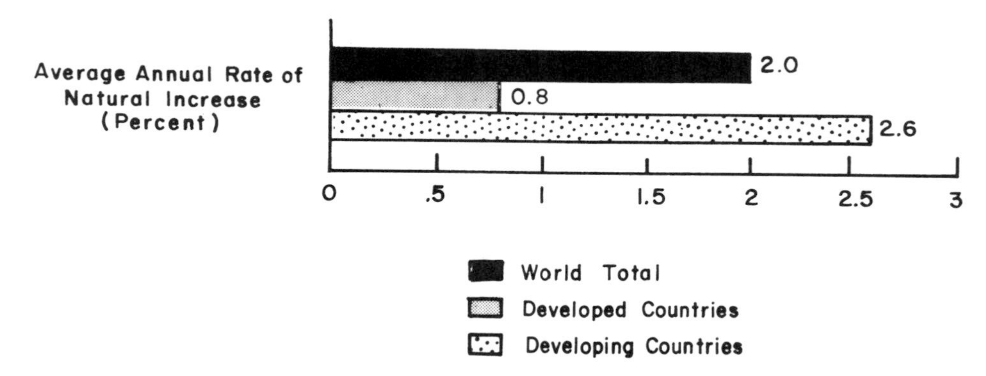

FIGURE 2-6. Current annual rate of population growth for world and for developed and developing countries. (Based on data from Frejka, T.: *The prospects for a stationary world population.* Sci. Am. 228:23, 1973.)

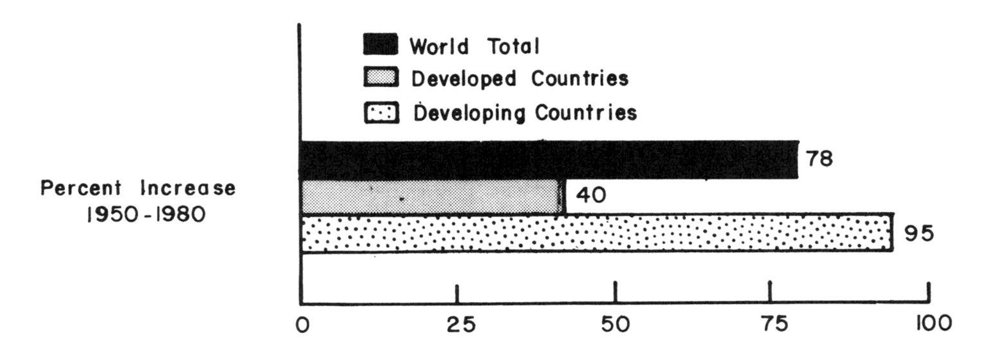

FIGURE 2-7. Projected percent increase in population for 1950 to 1980 for world and for developed and developing countries. (Based on data from Nortman, D.: *Population and family planning: a factbook*. Reports on Population/Family Planning. September 1973, p. 19.)

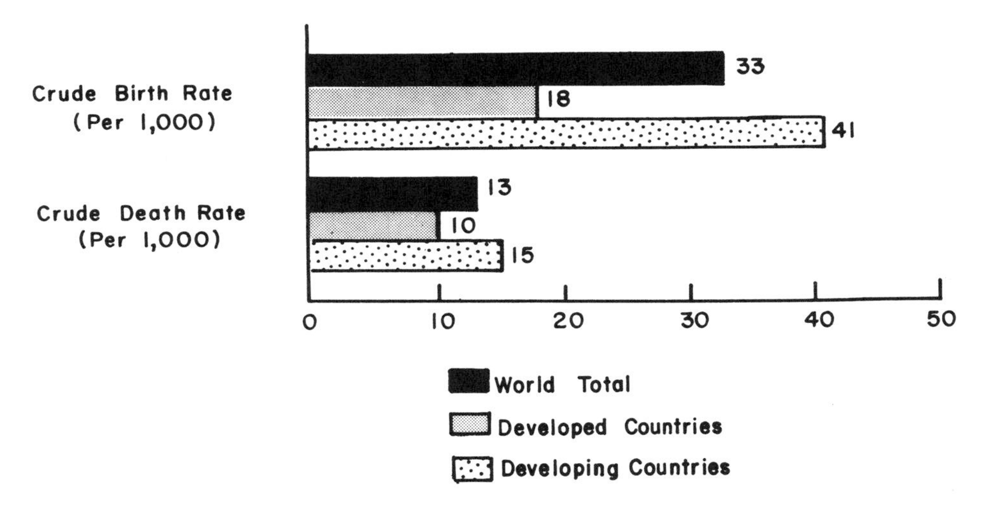

FIGURE 2-8. Crude birth and death rates per 1000 population for world and for developed and developing countries. (Based on data from Frejka, T.: *The prospects for a stationary world population*. Sci. Am. 228:23, 1973.)

The decrease in mortality has resulted in part from improvement in health care, sanitation, nutrition and changes in social, economic and political situations. The recent increased sophistication and expectation of the consumer must be viewed as an integral part of the social change also.

Developed and Developing Countries

As we will illustrate in the following sections, the rates of growth in developing countries (2.5 percent yearly, doubling time 28 years) exceeds that in developed countries (1.0 percent yearly, doubling time 70 years). The differences between the "haves" and "have nots" are increasing.

For instance, note the total population of the world and developed and developing countries in Figure 2-5 and compare this with the current annual population growth (Fig. 2-6) and a projected percent increase (Fig. 2-7). Observe the increase in crude birth and death rates in developing countries (Fig. 2-8) and observe fer-

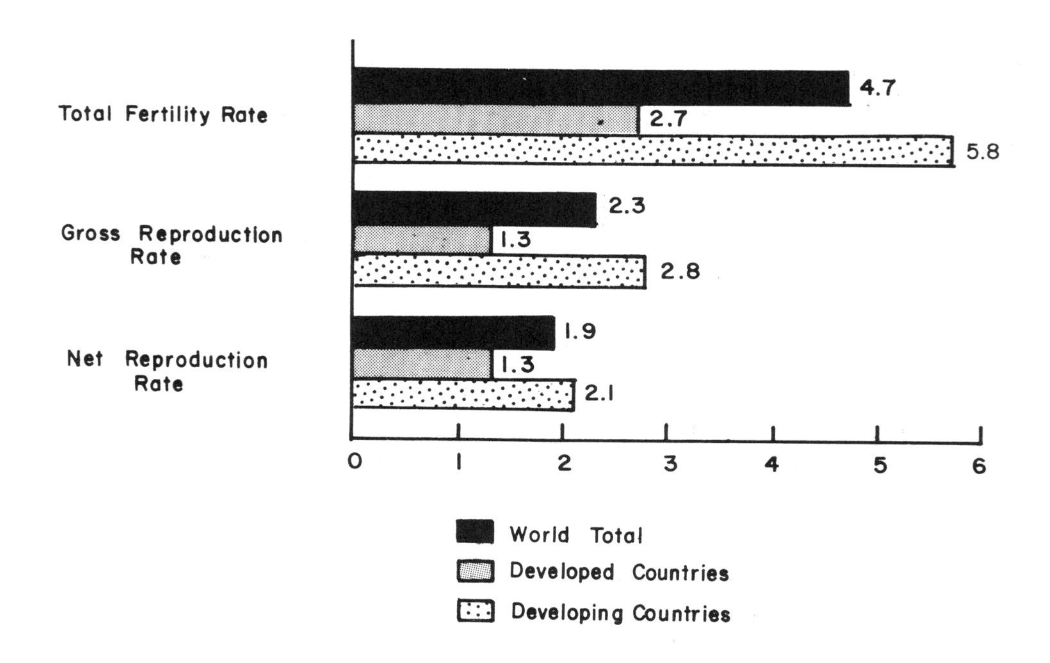

FIGURE 2-9. Current fertility patterns for world and for developed and developing countries. (Based on data from Frejka, T.: *The prospects for a stationary world population.* Sci. Am. 228:23, 1973.)

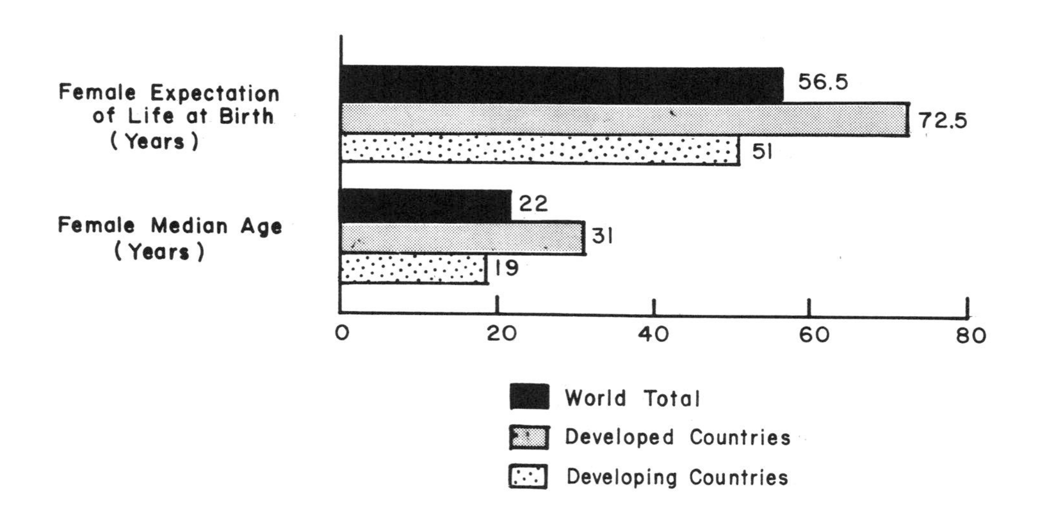

FIGURE 2-10. Demographic characteristics of females for world and for developed and developing countries. (Based on data from Frejka, T.: *The prospects for a stationary world population.* Sci. Am. 228:23, 1973.)

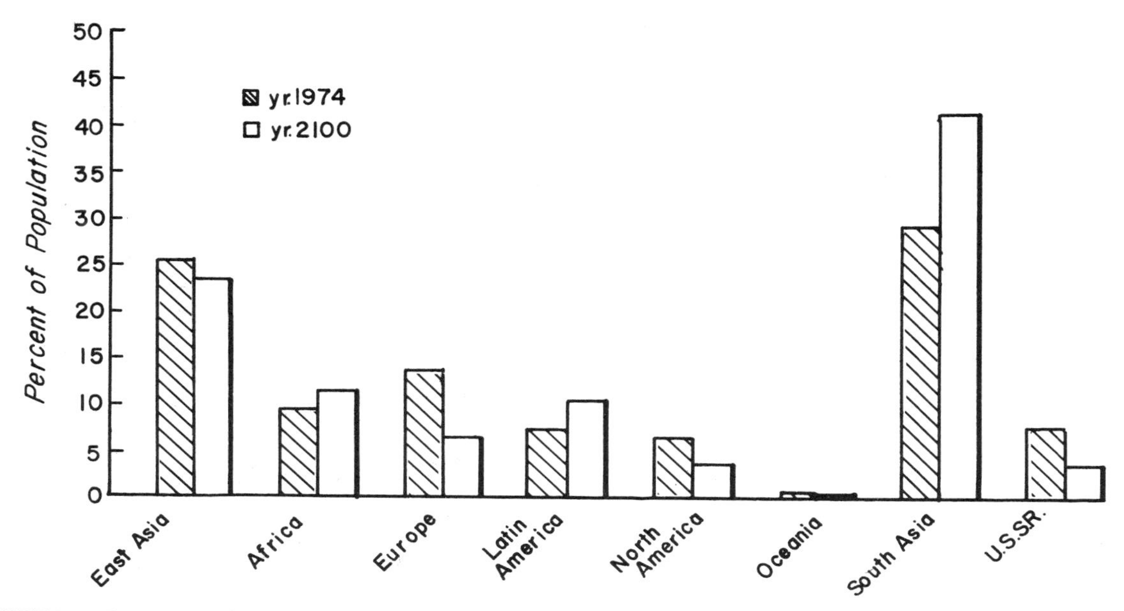

FIGURE 2-11. Percentage change in geographic distribution of world population 1974 to 2100. (Modified from Frejka, T.: *The prospects for a stationary world population,* Sci. Am., 228:23, 1973.)

tility rates in developed and developing countries (Fig. 2-9). Note in Figure 2-10 that, although the life expectancy is less for women living in developing countries, a higher percentage of these women are in their most fecund years.

Currently, 70 percent of the world's population lives in developing countries. This will increase to 80 percent by the year 2000. This change in the world's population distribution depicted in Figure 2-11 will have profound implications in the future.

Fertility Patterns

Study of the demographic features presented in Figure 2-11 and mentioned previously reemphasizes the fact that the present rapid growth of developing countries is due mainly to a marked difference in *fertility patterns,* there being little difference in mortality between developed and underdeveloped countries. Also of interest is the realization that most countries with high fertility (a crude birth rate greater than 30) are in the tropical zone between the Tropic of Cancer and the Tropic of Capricorn. Study of Table 2-1 will reveal this.

Furthermore, developed countries with currently low fertility rates have also shown declines in their crude birth rates over the past two decades (Table 2-2).

Table 2-1. Distribution of countries with respect to crude birth rates.

Crude Birth Rate	*Country*
55	Nigeria
50	Ethiopia, Kenya, Morocco, Sudan, Afghanistan, Iraq
	Algeria, Uganda
	Tanzania, Bangladesh, Iran, Pakistan
45	Columbia, Nepal, Philippines
	Zaire, Peru
	Mexico, South Vietnam
	Indonesia
	Brazil, Venezuela, Thailand
40	South Africa, Burma
	India, North Korea
	Egypt, North Vietnam, Turkey
35	
	China, West Malaysia
30	Chile, Sri Lanka
	South Korea
25	
	Taiwan
	Argentina
20	Romania, Australia
	Japan, Spain
	Soviet Union, Yugoslavia
	France, Poland
	Canada, Czechoslovakia, Italy, Netherlands
15	United States, Hungary, United Kingdom
	East Germany
	West Germany
10	

Reprinted by permission of The Population Council from Berelson, B.: *World population status report 1974*. Reports on Population Family Planning, No. 15, January 1974, p. 7.

Table 2-2. Crude birth rates—Europe and countries of European origin.

Country	*1960*	*1970*	*Percentage decline*
Austria	17.9	12.9	28
Belgium	16.9	13.5	20
Canada	26.7	15.5	42
Czechoslovakia	15.9	15.9	0
Democratic Republic of Germany	17.2	13.9	19
Finland	18.5	12.0	35
France	17.9	16.5	8
Germany (Federal Republic)	17.8	10.2	43
Greece	18.9	15.9	16
Hungary	14.7	14.7	0
Ireland	21.4	22.6	(+6)
Italy	18.3	16.1	12
Luxembourg	16.0	10.7	33
Netherlands	20.8	14.7	29
New Zealand	26.5	22.1	17
Norway	17.3	15.4	11
Poland	22.6	16.8	26
Portugal	24.2	21.2	12
Romania	19.1	21.1	(+10)
Spain	21.8	19.4	11
Sweden	13.7	13.3	3
Switzerland	17.6	13.8	22
United Kingdom	17.5	13.8	21
United States	23.7	15.9	33
USSR	24.9	17.4	30
Yugoslavia	23.5	17.8	24

Reprinted by permission of The Population Council from *Annual Report 1973*.

High Dependency Ratios

One of the major differences in the population characteristics of the developed and underdeveloped countries is the high dependency ratio in underdeveloped countries. This results in a large number of young people, many entering their reproductive years, thus providing a sustained momentum for population growth. Even with a net reproduction rate of 1.0 (replacement fertility), the population of these countries would still continue to grow well into the next century, since these young people would produce more births than the total population produces deaths. Until the age structure stabilizes, the population of a given area cannot be expected to reach a stable equilibrium. The fact that *approximately 50 percent of the population of the developing world is made up of people under 15 years of age* presents a formidable challenge to those who wish to achieve a stationary population in the near future.

As can be seen from a study of the population pyramid (Fig. 2-12), decreasing rates of growth in developing countries are reflected by an aging of the population.

Social and Economic Results

In addition to the perpetration of the vicious cycle of reproduction, the alarming growth of the population in the developing areas threatens to nullify the effects of economic expansion by dissipating this growth over more people, thereby reducing per capita gains. Presently, nearly two thirds of the world subsists on substandard incomes (Fig. 2-13). The gap between the "haves" and the "have nots," unfortunately, is widening.

Not only do developed countries have a

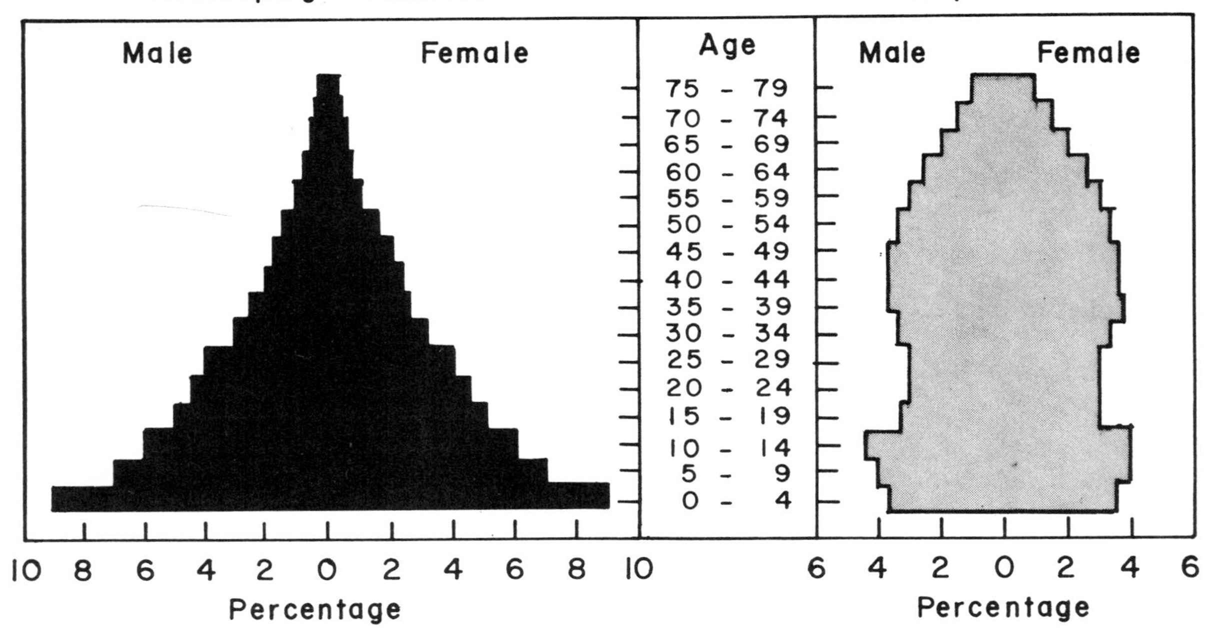

FIGURE 2-12. Population pyramid.

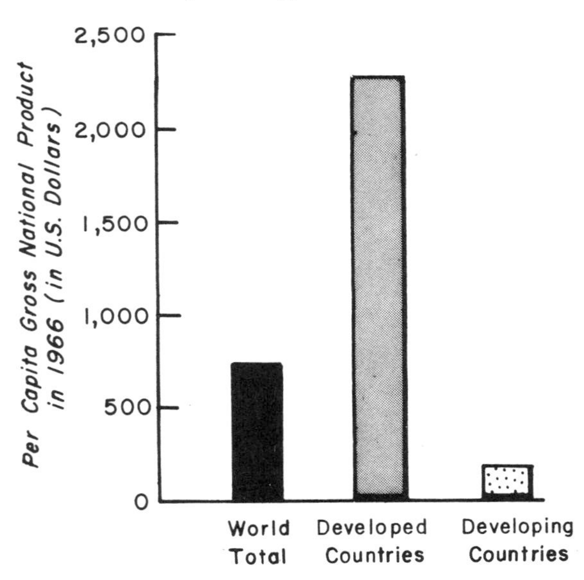

FIGURE 2-13. Per capita gross national product for the world and for developed and developing countries. (Based on data from Nortman, D.: *Population and family planning: a factbook*. Reports on Population/Family Planning. September 1973, p. 22.)

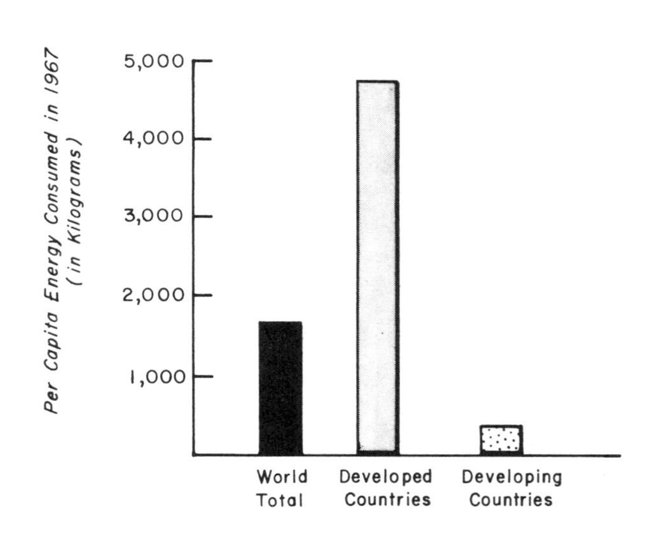

FIGURE 2-14. Per capita energy consumption for the world and for developed and developing countries. (Based on data from Nortman, D.: *Population and family planning: a factbook*. Reports on Population/Family Planning. September 1973, p. 22.)

higher per capita gross national product but they also consume more of the world's supply of energy than developing countries (Fig. 2-14).

A careful look at some of the social and economic factors resulting from the rapid population growth rate reveals that

1. In developing countries, in addition to a lower per capita income, there is an unequal distribution of this income. For example, in Peru only 10 percent of the population earns 55 percent of the national income.
2. The percentage increase in per capita food production has been dissipated over so many heads that progress becomes negligible (Fig. 2-15).
3. As a population doubles, that country simultaneously needs to double its economic output, housing, schools, health facilities, services and agriculture. The capital, which should be reserved for saving and development, must be utilized just to maintain the status quo. In developing countries of Asia, for example, an increase of over 50 percent in school facilities is required in the next 15 years just to maintain enrollment ratios in primary schools.[3]
4. Although more people are becoming educated, because of the growth in population, the number of illiterates is also rising. Therefore, even though world illiteracy fell from 40 to 35 percent over the past decade, the number of illiterates rose by 50 million.
5. The 77 percent increase from 1963 to 1971 in the gross national product of developing countries is deceiving since this gain was from a base level that was exceptionally low to begin with.
6. Developing countries tend to have a

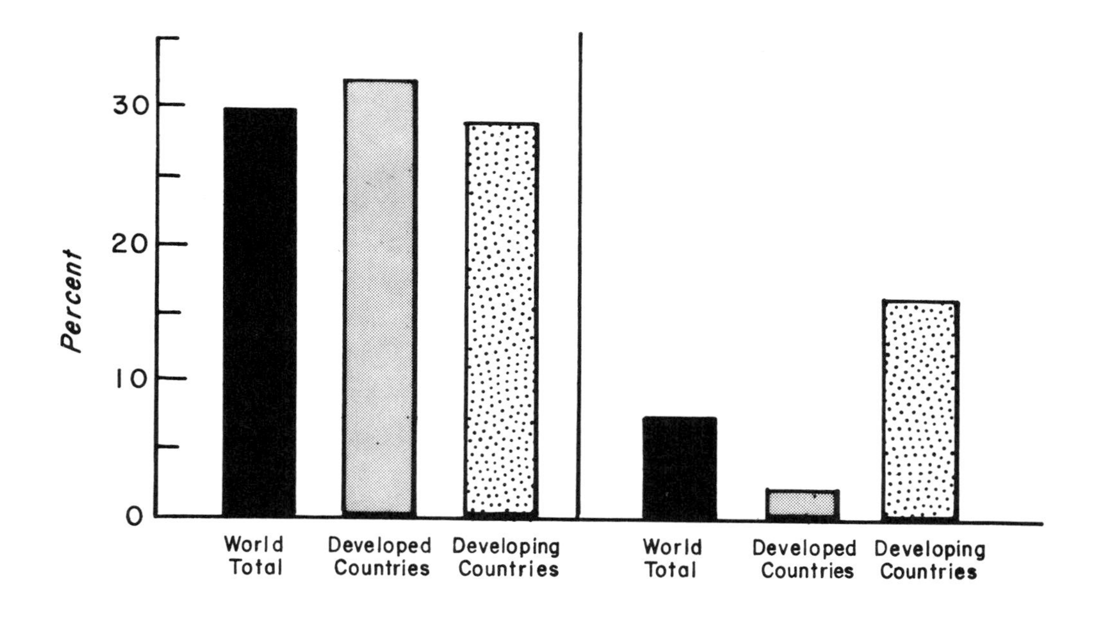

FIGURE 2-15. Percent increase in food production in ten-year period (1969–1971 average over 1959–1961 average). (Adapted from Nortman, D.: *Population and family planning: a factbook*. Reports on Population/Family Planning. September 1973, p. 22.)

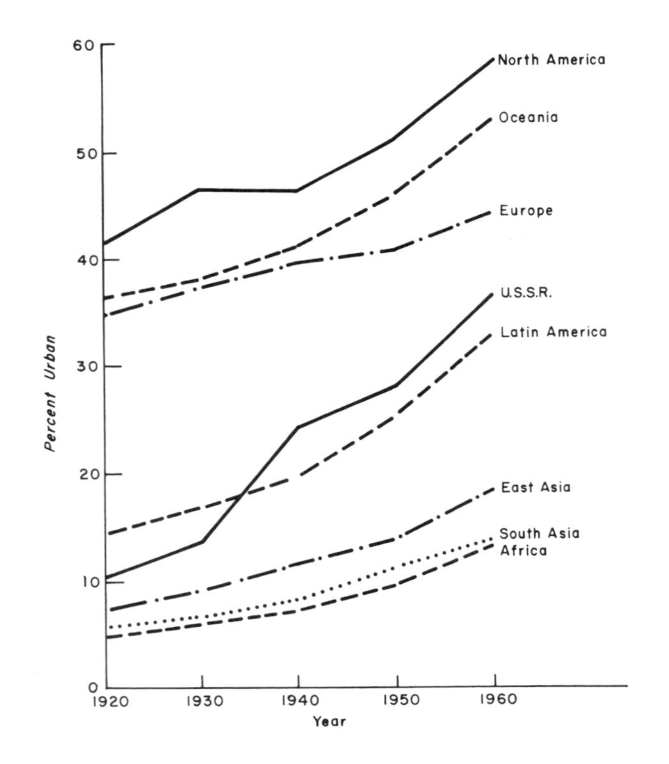

FIGURE 2-16. Growth of urban population by major areas, 1920 to 1960. (Adapted from Population Bulletin 29:15, 1974.)

higher proportion of their labor force in agriculture and a lower proportion in industrial activity and services.

7. Developing countries depend on a few primary commodities with a resultant instability in their export items, e.g., Cuba—sugar, Uruguay—meat.

8. Developing countries generally tend to be less urbanized with less than 50 percent of their populations living in cities[4] (Fig. 2-16). With technological development, however, internal migration will effect a large growth in the cities, which, subsequently, will cause demands for employment and services.

In contrast to the eight considerations above, declining fertility stabilizes and promotes economic growth. This effect is shown diagrammatically in Figure 2-17. There is a 15-year lag between the decline in fertility and the subsequent events shown in the figure.

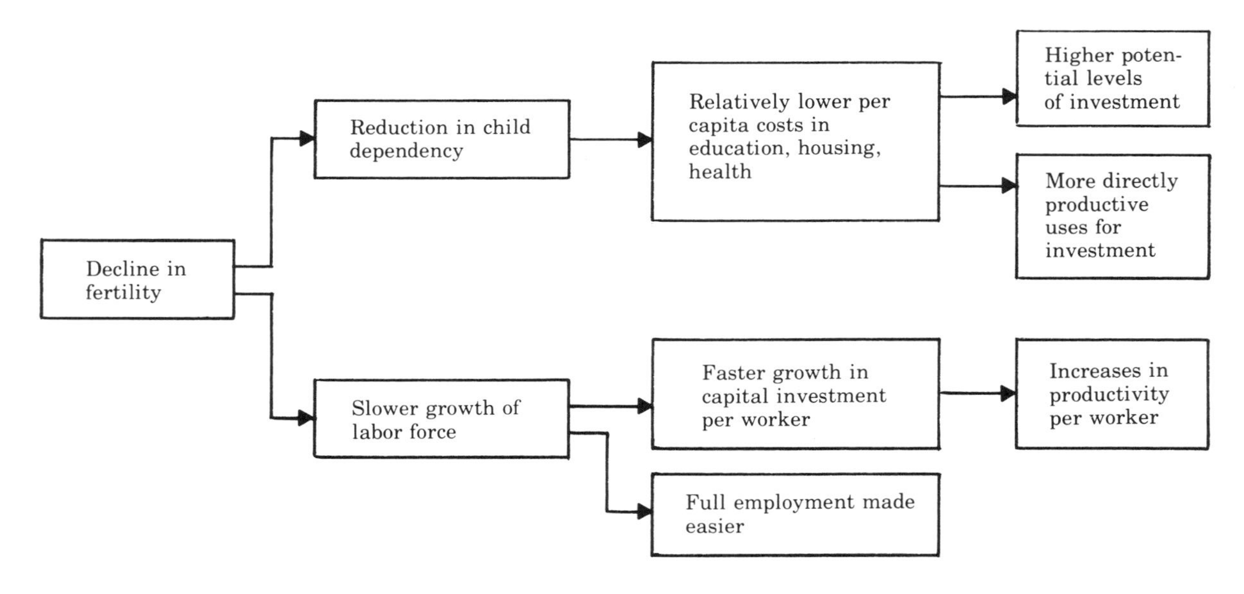

FIGURE 2-17. Some relationships between declining fertility and economic growth. (Adapted from Stamper, B. M.: *Population policy in development planning*. Reports on Population/Family Planning. No. 13, May 1973, p. 3.)

SOLUTIONS FOR LIMITING POPULATION GROWTH

There is no simple solution to the difficulties created by population growth. Any attempt to effect a solution is fraught with complications. However, we can begin with three basic ways to limit population growth:

1. increase the death rate—obviously an unrealistic solution.
2. emigration—essentially a form of *relocation* rather than a solution. This has been successful in 19th century Ireland and present day Barbados. However, emigration tends to deplete a nation of its work force and to increase its dependency ratio.
3. decrease the birth rate—this can be achieved by:

coercive or involuntary methods
social change
- education
- increasing the age of marriage (e.g., China)
- increasing the women in the labor force (e.g., Eastern Europe)
- industrialization
- urbanization

antinatalist incentives
- elimination of maternity benefits
- elimination of taxation benefits
- benefits to those restricting reproduction
- legalization of prostitution and homosexuality

voluntary measures
- adequate clinical facilities
- research in improved modalities

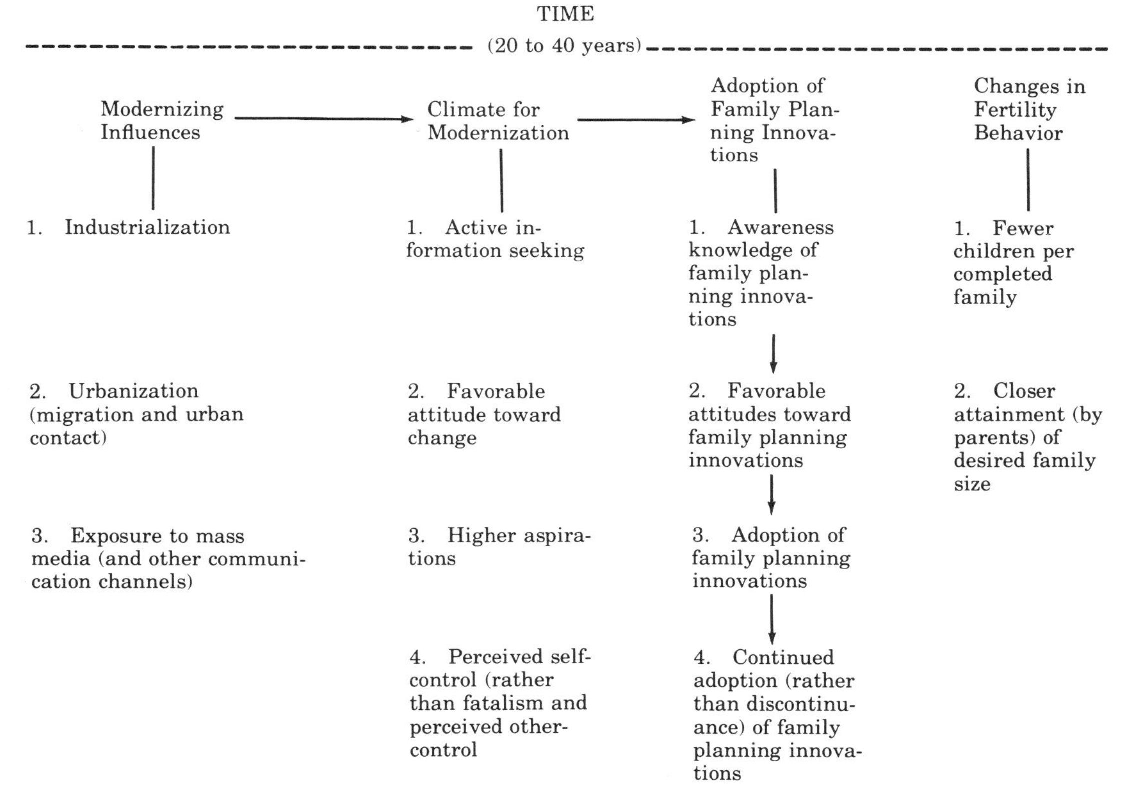

FIGURE 2-18. Schema of a long-range approach to fertility behavior changes through creation of a climate for modernization. (Adapted from Rogers, E.: *Communication Strategies for Family Planning*. The Free Press, Macmillan, London, 1973, p. 274.)

easy accessibility to family planning facilities and measures
public information

Of the three, the third method offers more realistic answers. Some of these, of course, are further complicated by moralistic, cultural or social mores. Think through each step in the third method. An example can be found in Figure 2-18.

Family Planning Services

A greater awareness on the part of governments of the need for family planning measures, brought about by numerous social demands and by the influence of a number of population groups, caused a sudden upturn in the mid-60's of family planning policies and/or programs in developing nations (Fig. 2-19). As a result of this upturn, by 1975 there were 54 developing nations with official policies and functioning programs to reduce fertility. An additional 32 nations provide family planning services for health and humanitarian reasons. Therefore, a total of 2.5 billion people, more than 90 percent of the developing world's population, live in countries that provide family planning services.

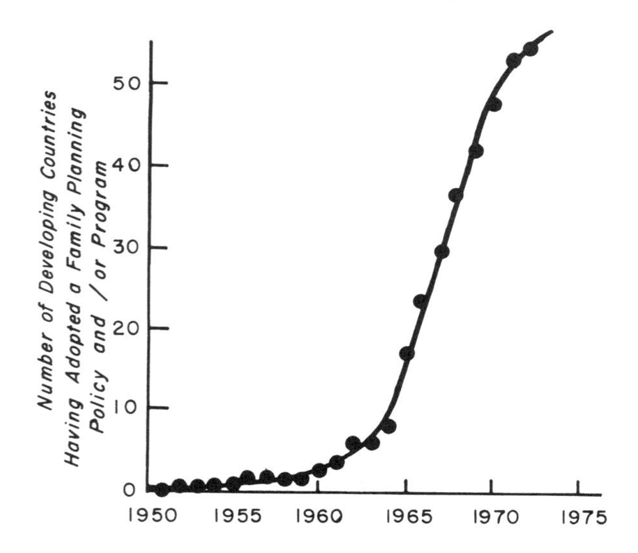

FIGURE 2-19. Growth in the number of countries with national family planning policies and/or programs since 1950. (Adapted from Rogers, E.: *Communication Strategies for Family Planning.* The Free Press, Macmillan, London, 1973, p. 7.)

These facts are deceiving. Despite the optimistic ring to the rise in family planning services, the following statistics by Mauldin[5] are eye-openers:

1. forty percent of these countries spend less than 10 cents per capita per year on family planning
2. four percent spend less than 20 cents per capita
3. family planning and population aid accounts for only 2 percent of all foreign assistance
4. less than 1 percent of national budgets is allocated to family planning

Although a decline has been noted in the crude birth rates in most countries where these demographic changes can be accurately measured (72/82 countries), the most rapid declines (Table 2-3) in general appear to be in countries with active family planning programs. Moreover, in several countries in which antecedent changes in social and economic development proceeded without active family planning programs, fertility decline did not occur (e.g., Kuwait, Brazil, Mexico).[6] One should keep in mind, however, Berelson's observation[7] that "in only about one-third of the most advanced countries in the world is modern fertility control widely available and practiced in the full array of technically and medically available means."

SUMMARY

The rapid growth of the world's population is alarming. At the present time, it takes only 35 years to double our population. Mortality rates, distribution of people, fertility patterns, and high dependency ratios are variables that affect the growth rate and/or are affected by them. Hardships on economy are caused by the population growth and the gap between the "haves" and "have nots" is widening. Three means of solutions are (1) increase the death rate, (2) emigration or relocation and (3) de-

Table 2-3. Changes in crude birth rates since 1960.

Countries	Crude Birth Rates 1960	1972	Percent change
Greenland	48.6	24.3 (1970)	50
St. Kitts-Nevis	43.3	22.7 (1970)	48
Hong Kong	36.0	19.4	46
Canada	26.7	15.7	41
Singapore	38.7	23.3	40
Barbados	33.6	20.4	39
Taiwan (China)	39.5	24.2	39
Granada	44.7	27.9 (1970)	38
Germany (Federal Republic)	17.8	11.4	36
Malta	26.1	16.8	36
Mauritius	38.5	25.0	35
United States	23.7	15.6	34
Trinidad and Tobago	37.9	25.1	34
St. Vincent	49.8	33.2 (1969)	34
Costa Rica	47.4	31.6	33
Reunion	44.0	29.5	33
Martinique	37.4	25.1	33
German Democratic Republic	17.2	11.7	32
Finland	18.5	12.7	31
Fiji	39.9	27.8	30
Iceland	28.0	19.7 (1971)	30
Union of Soviet Socialist Republics	24.9	18.0	28
Brunai	48.9	35.4	28
Luxembourg	16.0	11.8	26
Bermuda	27.3	20.7 (1970)	24
Guadeloupe	38.4	29.4	23
Poland	22.6	17.4	23
Netherlands	20.8	16.1	23
Dominica	47.0	36.4 (1969)	23
Yugoslavia	23.5	18.2	23
Cape Verde Islands	44.8	34.9 (1971)	22
Egypt	43.0	34.6 (1971)	20
Austria	17.9	13.8	23
Albania	43.4	35.3 (1969)	19
Belgium	16.9	13.8	18
Sri Lanka	36.6	29.9 (1971)	18
Switzerland	17.6	14.4	18
El Salvador	49.5	40.7	18
New Zealand	26.5	21.8	18
Jamaica	42.0	34.6	18
Canal Zone, Panama	18.2	15.0	18
Cook Islands	49.5	40.9 (1968)	17
Chile	35.7	29.6 (1970)	17
American Samoa	42.9	35.8 (1971)	17
Portugal	24.2	20.3	16
Greece	18.9	15.9 (1971)	16
Faeroe Islands	22.5	19.0 (1971)	16
Antigua	34.1	30.4 (1965)	11
United Kingdom	17.5	14.9	15
Guatemala	48.9	41.7 (1971)	15
Bulgaria	17.8	15.3	14
Guyana	42.2	36.3 (1968)	14
West Malaysia	40.9	33.6	18
Panama	41.0	35.6	13
Tunisia	46.8	41.0 (1969)	12
Channel Islands	16.5	14.6 (1971)	12

Table 2-3. Changes in crude birth rates since 1960 (*Continued*)

Countries	*1960*	*1972*	*Percent change*
Spain	21.8	19.4	11
Surinam	45.6	40.9 (1966)	10
Australia	22.4	20.5	8
Denmark	16.6	15.2	8
Italy	18.3	16.3	11
Guam	36.7	34.1 (1971)	7
Gibraltar	23.7	22.1 (1971)	7
Puerto Rico	32.3	25.6 (1971)	21
France	17.9	16.9	6
Uruguay	23.9	22.6 (1971)	5
Algeria	48.2	46.0 (1968)	5
Sao Tome and Principe	46.1	44.0 (1971)	5
Norway	17.3	16.6	4
Argentina	22.7	21.9 (1968)	4
Mexico	44.6	43.4	3
St. Lucia	49.3	48.9 (1970)	1
Israel	26.9	26.9	No change
Hungary	14.7	14.7	No change
Sweden	13.7	13.8	1
Romania	19.1	19.6 (1971)	3
Jordan	46.3	47.8 (1966)	3
Czechoslovakia	15.9	16.5 (1971)	4
Ireland	21.4	22.4	5
Isle of Man	14.1	15.0	6
Japan	17.2	19.3 (1971)	12
U.S. Virgin Islands	36.6	49.5 (1970)	35

Adapted from Ravenholt, R.T., and Chao, J.: *World Fertility Trends, 1974*. Population Report, Series J, No. 2, 1974, p. 21.

crease the birth rate. Vigorous family planning services are necessary to affect the crude birth rate and, thus, to begin the procedure of establishing equilibrium. The longer the problem is ignored the more difficult will be the solution.

QUESTIONS

1. What problems must a country with a high dependency ratio deal with?
2. What are the disadvantages of rapid population growth in a developing country?
3. What programs would you suggest to reduce a developing country's fertility?

REFERENCES

1. Hauser, P. M.: *World population growth*. In Hauser, P.M.: *The Population Dilemma*. Prentice-Hall, Inc. Englewood Cliffs NJ, 1969, p. 12.
2. As reported in *People*. International Planned Parenthood Federation, Vol. 1, No. 5, 1975.
3. Population Studies Series, United Nations, No. 11.
4. Berelson, B.: *World Population: Status Report 1974*. Reports on Population/Family Planning. January 1974.
5. Mauldin, W. P.: *Family planning programs and fertility declines in developing countries*. Fam. Plann. Perspect. 7:32, 1975.
6. Ravenholt, R. T., and Chao, J.: *World fertility trends, 1974*. Population Report, Series J, No. 2, 1974, p. 21.
7. Berelson, B. (ed.): *Population Policy in Developed Countries*. McGraw-Hill Book Co., New York, 1974.

BIBLIOGRAPHY

Berelson, B.: Report of the President. The Population Council Annual Report, 1971.

Bhagwati, J.: *The Economics of Underdeveloped Countries*. World University Library, McGraw-Hill Book Co., New York, 1966.

Frejka, T.: *The prospects for a stationary world population*. Sci. Am. 228:15, 1973.

Hauser, P. M.: *The Population Dilemma,* Prentice-Hall, Inc., Englewood Cliffs NJ, 1969.

Nortman, D.: *Population and family planning: a factbook*. Reports on Population/Family Planning. January 1974.

3 Population Problems in the United States

The United States today is characterized by low population density, considerable open space, a declining birthrate, movement out of the central cities—but that does not eliminate the concern about population. This country, or any country, always has a "population problem" in the sense of achieving a proper balance between size, growth, and distribution on the one hand, and, on the other, the quality of life to which every person in this country aspires.[1]

The belief that the United States does not have a population problem must be attributed to wishful thinking; it is not based on factual

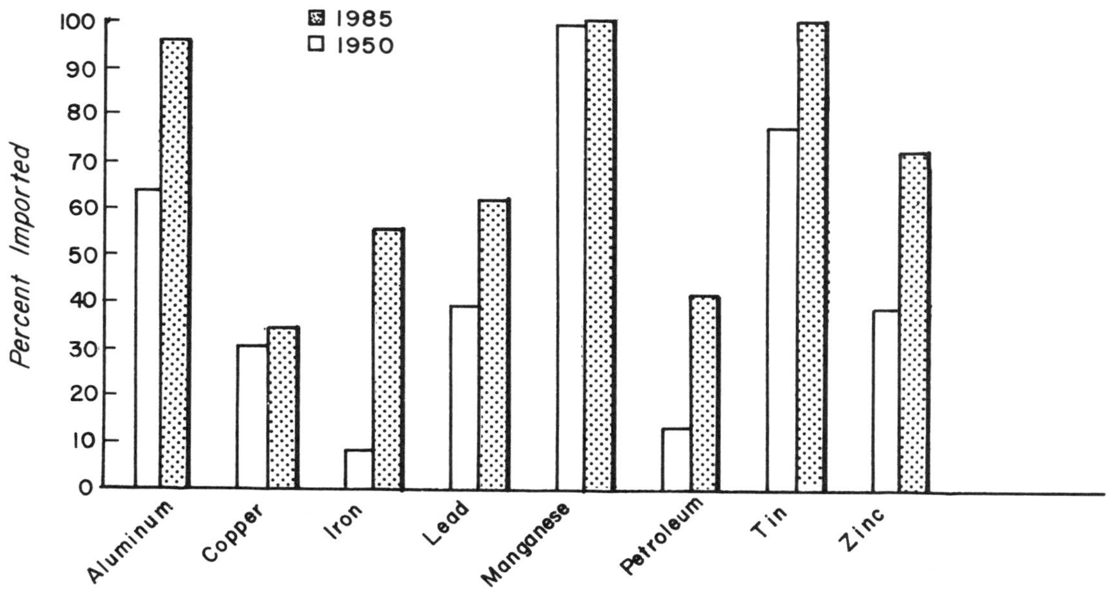

FIGURE 3-1. Increasing dependency of the U.S. on imports. (Based on data from U.S. Department of Health, Education, and Welfare (NIH) publication, July 5, 1975.)

information. In fact, regardless of seemingly endless resources and open space, the growth of the population of the United States presents serious problems both on a national and an international level.

PROBLEMS AND IMPLICATIONS

Despite the fact that the United States has only 6 percent of the world's population, it yearly consumes between 35 and 40 percent of the world's products. For instance, it consumes 57 percent of the natural gas, 42 percent of the world's silver, 36 percent of the aluminum, and 32 percent of the petroleum. Because of its economy, it is able to outbid underdeveloped countries for these materials; this power further increases prices and lowers supplies.

Moreover, the United States is becoming increasingly dependent on foreign imports. Compare the percent of various raw materials imported in 1950 with projected imports in 1985, as shown in Figure 3-1.

Despite the fact that considerable financial resources buy material objects on the world market, monetary measures do not solve all the population pressures at home. These pressures are noticeable in urban congestion, racial unrest, unemployment, poverty, air and water pollution, and the rise of crime. They also affect agricultural resources, outdoor recreational areas, and highway systems. Consider the reasons for population growth that cause these pressures.

Trends in Fertility

The highest birth rate (55/1000) in United States history was established in colonial America. This was followed by a steady decline until 1940 when a low of 18/1000 was reached following the depression. The rate rose again during World War II (27/1000) and this rate persisted until 1957. Of interest is the fact that this "baby boom" was a result not of larger families but of more families of moderate size. The rate and number of births then decreased until the late 60's when there was an interim spurt in the number of births (Table 3-1), caused by an increase in the number of females of childbearing age, a result of the baby boom.

Table 3-1. United States birth patterns.

Year	*Total births* $\times 10^3$	*Birth rate*
1960	4258	23.7
1962	4167	22.4
1963	4098	21.7
1964	4027	21.0
1965	3760	19.4
1966	3606	18.4
1967	3521	17.8
1968	3502	17.5
1969	3600	17.8
1970*	3731	18.4
1971*	3556	17.2
1972*	3258	15.6
1973*	3137	14.9
1974†	3166	15.0

Modified from *Live Births, Deaths, Marriages and Divorces: 1910–1974*. Statistical Abstract of the U.S. Department of Commerce, Bureau of Census, ed. 96, 1975, p. 51.

*Excludes nonresident aliens

†Preliminary

The United States is currently experiencing the lowest fertility in its history, and the rates continue to fall. For example:

1. There were 9 percent fewer births in 1972 than in 1971 despite a 3 percent increase in women of childbearing age.
2. During this one-year period the crude birth rate fell 10 percent (15.5) and the fertility rate fell 11 percent (to 72.7).
3. In 1974 the birth rate fell even lower (15.0), and the fertility rate fell to 68.4. This is nearly half the rate of 1957 (122.7).
4. Birth expectancy is decreasing, e.g., for ages 18 to 24, in 1973 there was an expectation of 2.3 births as compared to an expectation of almost three births in 1967; and, for ages 25 to 29, there was a 1973 expectation of 2.4 births as compared to a 1967 expectation of three births per female.[2]

Despite this decline in fertility, the population of the United States continues to grow at approximately 1 percent per year. This is due in part to the fact that there are a large number of females within their reproductive years. Moreover, the National Center for Health Statistics has pointed out that the

Table 3-2. Demographic perspective of 20th century United States*

	Circa 1900	*Circa 1970*
Population	76 million	205 million
Life expectancy	47 years	70 years
Median age	23 years	28 years
Births per 1000 population	32	18
Deaths per 1000 population	17	9
Immigrants per 1000 population	8	2
Annual growth	1¾ million	2¼ million
Growth rate	2.3%	1.1%

Data from *Report of the Commission on Population Growth and the American Future.* March, 1972. Adapted from Stud. Fam. Plann., Vol. 3, No. 5, 1972.

*Figures are rounded off.

number of women of reproductive age in the United States will increase by an additional 12 percent by 1980. There also has been a decline in the death rate to 9.1 per 1000 population and in infant mortality to 16.5 per 1000 live births, reaching this rate in 1974. This is the lowest rate ever recorded in the United States.[3] A demographic presentation of the United States population is given in Table 3-2.

At the present rate of growth, nearly 1.5 million people are added per year to the United States population; this is enough people to occupy a city the size of Philadelphia. Diagrammatic comparison of the projected future size of the population at replacement fertility, which is two children per family, and with a three-child family, is shown in Figure 3-2. One can appreciate that, even if replacement fertility were practiced, the concept of zero population growth cannot be achieved until after the turn of the century.

Implications. Among the implications of the present low birth rate is an anticipated lower proportion of younger people (less than 20 years of age) and a higher percentage of adults 20 to 34 in the next decade (Fig. 3-3). This is the reverse of the situation created by the baby boom of the 50's.

The present changes in population growth rates in the United States are dependent basically on fluctuating birth rates because mortality and immigration rates are stable. These periodic fluctuations place obvious strain on economic and social institutions such as schools, hospitals and the labor market.

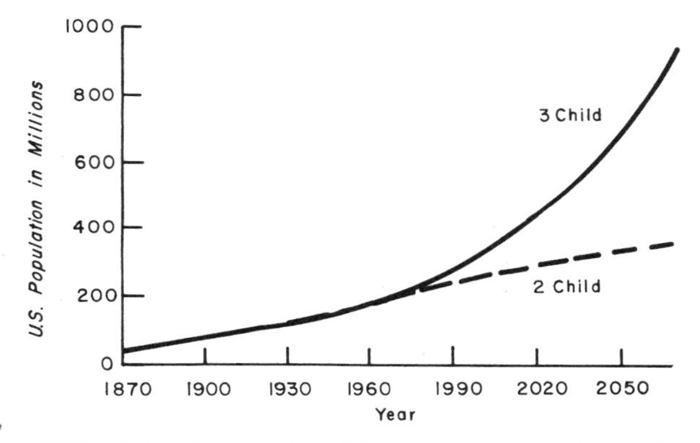

FIGURE 3-2. Future size of U.S. population with 3-child vs. 2-child family. (Adapted from *The Report of the Commission on Population Growth and the American Future,* March 1972.)

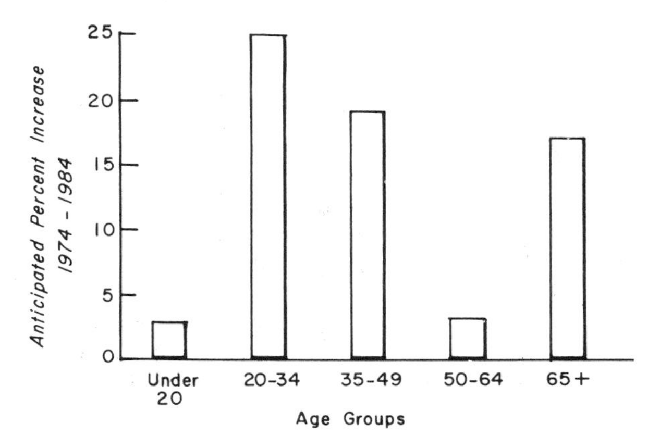

FIGURE 3-3. Anticipated percentage increase in population of age groups in the U.S. 1974 to 1984. (Based on data from U.S. News & World Report, February 25, 1974, pp. 44 and 45.)

FACTORS AFFECTING FERTILITY

There are many factors that affect fertility in the United States. Several examples are given here for illustrative purposes. Each statement should be thoughtfully considered because this is merely a perusal of the subject.

Marriage

The age at marriage in the United States is the lowest of any developed country in the world and has changed little during the past decade:

	1960	1970
females	20.2	20.8
males	22.5	23.2

From 1958 to 1973 there has been an increase in the number of marriages and in the marriage rate in the United States. However, 1974 marked the first annual decrease since 1958, a decrease of 3 percent. This has been complimented by an increase in the divorce rate (e.g., for each two American couples marrying today one can expect a divorce) and an increase in the pregnancy rate of unmarried women (see Chapters 5 and 6).

Statistically, the younger a woman is at the time of marriage, the higher will be her completed fertility (Fig. 3-4). However, the percentage of women remaining single until they are between 20 and 24 years old has increased from 28 percent in 1960 to 40 percent in 1974.[4]

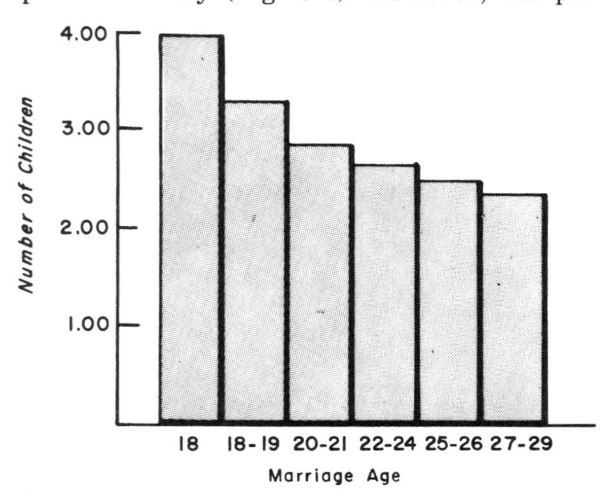

FIGURE 3-4. Average number of children ever born to U.S. mothers who had completed their childbearing by 1960. (Adapted from The Population Council Annual Report, 1971.)

Minorities

If all other variables that influence fertility in the United States were excluded, one would find a higher rate of fertility among the nonwhite population (Fig. 3-5). Although the rate is higher, it is decreasing (Fig. 3-6).

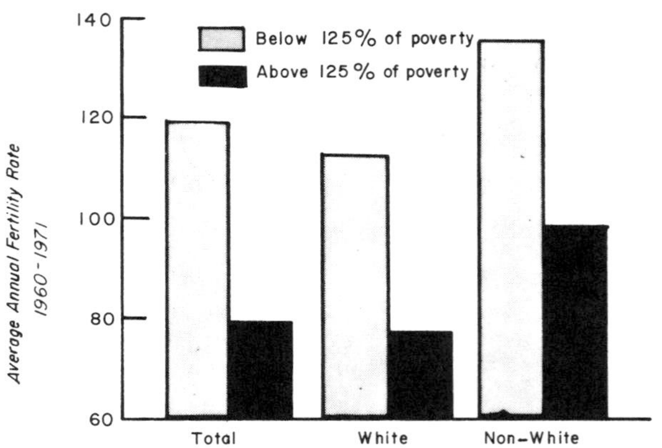

FIGURE 3-5. Average annual fertility rates for race and poverty status, 1966 to 1971. (Based on data from Westoff, C. F.: *The modernization of U.S. contraceptive practice.* Fam. Plann. Perspect. 4:9, 1972.)

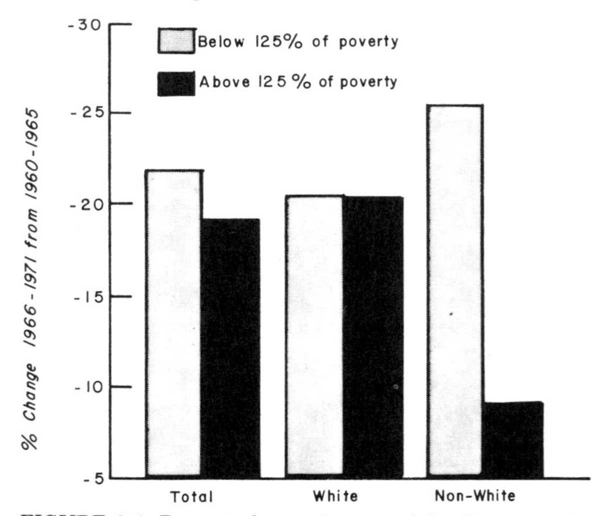

FIGURE 3-6. Percent change in annual fertility rates for race and poverty status, 1966 to 1971. (Based on data from Westoff, C. F.: *The modernization of U.S. contraceptive practice.* Fam. Plann. Perspect. 4:9, 1972.)

The result of this differential fertility in Figure 3-6 will be a change in the racial make-up of the population. Between 1955 and 1965 the non-white population increased 27 percent while the total United States population increased 17 percent. In 1966, 11 percent of the population was non-white and this will rise to 14 percent by 1985. Notice the sharp changes in Table 3-3.

Table 3-3. Projected percentage increase of population from 1966 to 1985.

	Non-white	*White*
Total population	68%	37%
Females, 15–44 yr.	65%	40%
Population under 15 yr.	60%	26%

Adapted from Hauser, P. M.: *The Population Dilemma.* Prentice-Hall, Inc., Englewood Cliffs NJ, 1969, p. 100.

Table 3-4. Projected percentage increase in labor force from 1966 to 1985.

Age of labor force	*Non-white*	*White*
15–64	75%	41%
15–44	124%	67%

Adapted from Hauser, P. M.: *The Population Dilemma.* Prentice-Hall, Inc., Englewood Cliffs NJ, 1969, p. 100.

The effects of changes in the relative size of various segments of the population will be noted in the racial make-up of schools and the labor force (Table 3-4) as well as in future fertility patterns.

If no children were born to black or Spanish-speaking parents in the 1960's, the United States population would be 4 percent smaller today. If there were no births to whites, the population would be 13 percent smaller.

Education

As can be seen in studying Figure 3-7, educational status of the female has an inverse effect on fertility.

Educational status has a compensatory effect on the marked difference in fertility for non-whites mentioned previously. Differential fertility in relation to income, occupational class

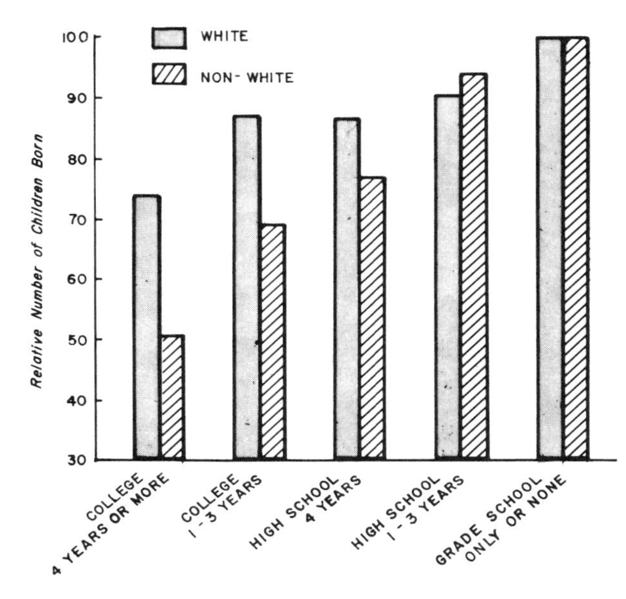

FIGURE 3-7. Relative number of children born to women in given age class with 100 as a base. (Data from Hauser, P. M.: *The Population Dilemma.* Prentice-Hall, Inc., Englewood Cliffs, NJ, 1969, p. 201.)

or education is much greater within our ethnic minorities than in the white population. For high school graduates as a whole, black fertility is lower.

Poverty

Fertility rates are directly related to poverty status (see Figs. 3-5 and 3-6).

During the second half of the 1960's, fertility declined more rapidly among poor and near-poor females than among those with incomes above the poverty level. This reflects in part the recent availability of family planning facilities for the poor.

Illegitimacy

The high rates of illegitimacy, especially among adolescents, is of great concern (see Chapter 6). The question often arises—Do welfare payments influence the illegitimacy rate? The answers are (1) there is no relationship between the number or rate of illegitimate births and the size of the welfare rolls (Aid to

Families of Dependent Children), and (2) illegitimacy continued to increase at the same rate in states that paid low benefits as in those that paid relatively high benefits.

Immigration

Each year 400,000 legal immigrants enter the United States. This is the highest immigration rate in the world. Since 1820 there have been 45 million immigrants. Between 1960 and 1970 about 16 percent of the total population growth of the United States was due to immigration.

More than 500,000 illegal immigrants enter the country each year, most of them poorly educated and with little resources. Many of them remain permanently. We must consider effects on the labor market and effects on social services.

To the advantage of the United States, most of the previously mentioned legal immigrants are among the better educated and more intrepid of the country of their origin. However, this circumstance then depletes a valuable resource of that developing country from which the immigrant comes, especially after that particular country has invested in the welfare and education of that individual.

Internal Migration

Shifts in population structure within a country can have strong social and economic implications. For instance,

Rural to urban—one sixth of the population lives in the metropolis from Boston to Washington.

Cores of urban centers to peripheries—phenomenon of suburbia.

Interior to coastal—approximately 53 percent of the population now lives in counties within 50 miles of the coasts.

Older metropolitan centers in the Northeast and Midwest to new centers in the South and West (Compare Fig. 3-8 and Table 3-5).

FIGURE 3-8. The sunbelt region is leading the nation in population growth.

Table 3-5. Population shift, 1970 to 1975.

Region and state	*Residential population (in thousands)*		*Percentage change*
	July 1, 1975	*April 1, 1970*	
United States, total	213,121	203,304	+4.8
Northeastern states	49,461	49,061	+0.8
North Central states	57,669	56,593	+1.9
The South	68,113	62,812	+8.4
The West	37,878	34,838	+8.7

Data from *Live Births, Deaths, Marriages and Divorces: 1910–1974.* Statistical Abstract of the U.S. Department of Commerce, Bureau of Census, ed. 96, 1975, p. 51.

In 1960 18 percent of all inner city residents were non-white. By 1985 this will rise to 31 percent. Recently, it has been noted, the trend of rural to urban migration has been reversed with subsequent increasing growth of rural areas and small towns.

SUMMARY

With 6 percent of the world's population, the United States consumes between 35 and 40 percent of the world's products. This fact and the growth of the population, approximately 1 percent yearly, presents serious problems nationally and internationally. Trends in the fertility of the United States and factors that influence or are influenced by these trends will have a profound effect on United States social and political institutions, life styles, economy and environment. The United States *does* have a population problem. Whether it can solve its problem remains to be seen.

QUESTIONS

1. How does a country like the United States with a low fertility rate contribute to the world's population pressures?
2. Why is the population of the United States continuing to grow?
3. What economic and social changes can you see as a result of a decreasing birth rate?
4. With immigration, what are the advantages and disadvantages to both the United States and the country of emmigration?

REFERENCES

1. Report of the Commission on Population Growth and the American Future, March 1972.
2. United States Department of Health, Education, and Welfare (NIH) Publication No. 75-781.
3. Monthly Vital Statistics Report. National Center for Health Statistics. Vol. 23, No. 12, Feb., 1975.
4. *New York Times,* July 6, 1975, p. 1.

BIBLIOGRAPHY

Cain, G.G., and Weininger, A.: *Economic determinants of fertility.* Demography 10(2):205, 1973.

Cutright, P.: *AFDC, Family allowances and illegitimacy.* Fam. Plann. Perspect. 2:4, 1970.

Frejka, T.: *The prospects for a stationary world population.* Sci. Am. 228:15, 1973.

Hauser, P.M.: *The Population Dilemma,* Prentice-Hall, Inc., Englewood Cliffs, NJ, 1969.

4 Population Stresses on the Family Unit

Previous chapters have dealt with overpopulation pressures on the world and individual nations. This chapter examines some of the stresses of family size on the family unit itself. The situation is reversed. One should now ask how size affects the family unit rather than how the family affects society. One must consider economic and social factors related to family size and further consider the effects of family size on maternal and child health, the effect of birth interval, and high-risk women of extreme reproductive ages.

ECONOMIC FACTORS IN THE LARGE FAMILY

It is expensive to have and raise a child, even in a one-child family. The total cost of a first child (with 1969 figures) is indicated in Table 4-1.

With the total cost of one child in mind, it is obvious that a large family puts stresses on the family budget. Only two examples of diminished per capita expenditure with increasing family size are shown here. Figure 4-1 shows the percentage of income (in three in-

Table 4-1. Total cost of a child (1969).

	*Discounted**	*Undiscounted**
Cost of giving birth	1,534	1,534
Cost of raising a child	17,576	32,830
Cost of a college education	1,244	5,560
Total direct cost		39,924
Opportunity cost for the average woman†		58,437
Total costs of a first child	59,627	98,361

From Ritchie, H., McIntosh, R., and McIntosh, S.: *Costs of Children*. Prepared for The President's Commission on Population Growth and the American Future, Washington, DC, 1972.

*Discounted and undiscounted costs—spending $1000 today costs more than spending $1000 over a 10-year period because of the nine years of potential interest on the latter. This fact is allowed for on money not spent in the first year. True costs are not accurately reflected in the undiscounted estimates, for these are simply accumulations of total outlays without regard to the year in which they must be made.

†Depending on the educational background of the mother, the opportunity costs (earnings foregone by not working) could be higher or lower.

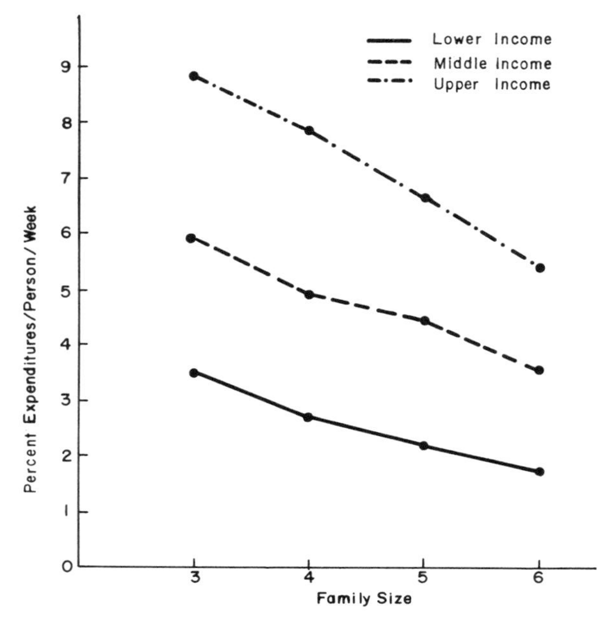

FIGURE 4-1. Calculated food expenditures per person per week by family size and income, United States, 1960 to 1961. (Based on data from Wray, J. D.: *Population pressure on families: family size and child spacing.* Reports on Population/Family Planning 9:447, 1971.)

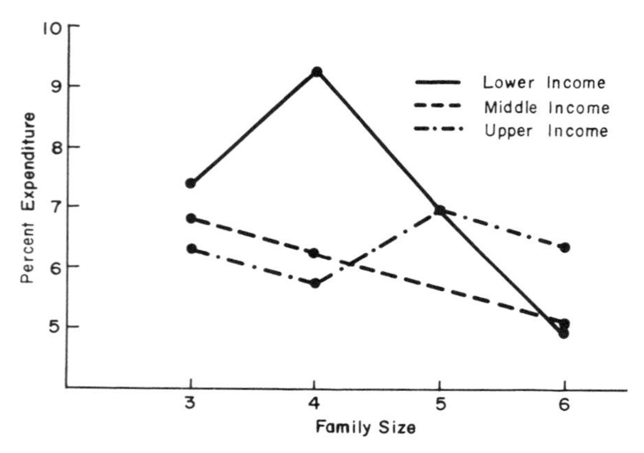

FIGURE 4-2. Expenditures for medical care as percent of total expenditures for current consumption by family size and income, United States, 1960 to 1961. (Based on data from Wray, J. D.: *Population pressure on families: family size and child spacing.* Reports on Population/Family Planning 9:449, 1971.)

come levels) spent on each person for food with increasing family size. The same study is made of medical care expenditures in Figure 4-2.

HEALTH CONSEQUENCES OF FAMILY SIZE

This decrease in expenditure for medical care, shown in Figure 4-2, occurs at the time the need for such care increases. As the size of the family increases, so do the health care problems. Some of the consequences of these factors are given in the following pages.

Health of the Children

The following statements indicate how family size influences the health care of the child.

1. There is an increased incidence of illness in children born to mothers of higher parity.
2. Nutrition, as measured by per capita monthly food purchases, is inversely related to family size. This is especially true in lower income families.
3. The neonatal, infant and childhood mortality rates increase directly in proportion to family size (Fig. 4-3). There is no regard to social class in the increase (Fig. 4-4). These mortality rates are generally most pronounced in mothers at the extremes of the reproductive cycle, also regardless of class.

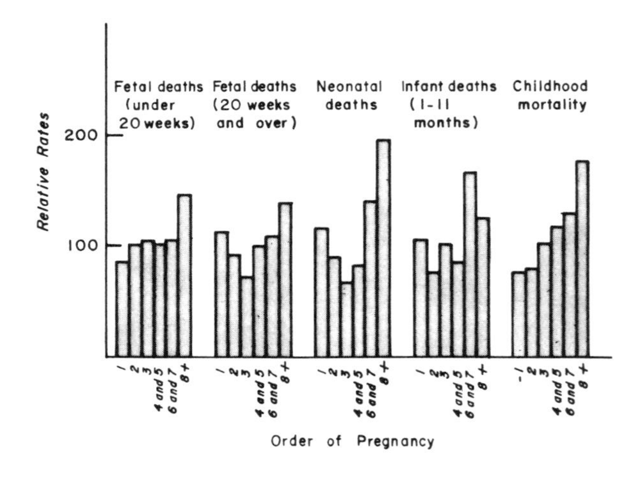

FIGURE 4-3. Variations in relative mortality rates with order of pregnancy from gestation to early childhood. (Adapted from Wray, J.D.: *Population pressure on families: family size and child spacing.* Reports on Population/Family Planning 9:418, 1971.)

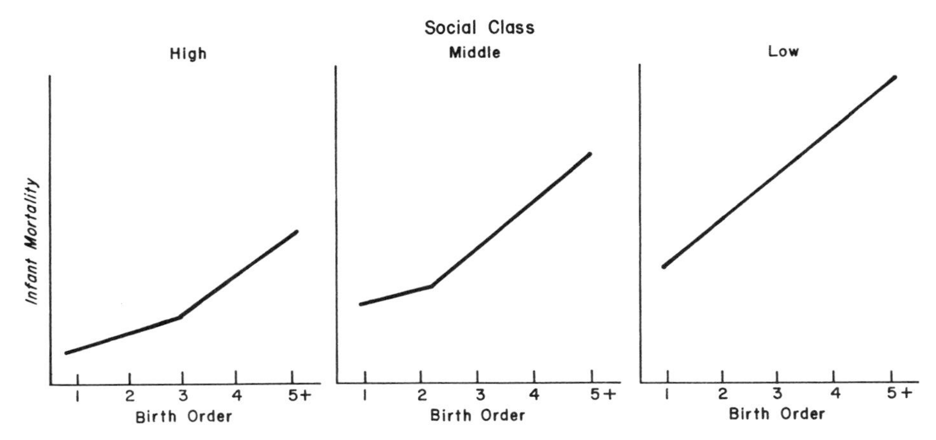

FIGURE 4-4. Risk of infant mortality (1 to 2 months) by social class and birth order (in England and Wales). (Adapted from Omran, A. R.: *Health benefits for mother and child.* World Health, January 1974, p. 64.)

4. Poor physical growth (height and weight) and later age of menarche has been noted in all siblings of larger families. This is less noticeable however by social class.[1]

5. Children from larger families score significantly lower in intelligence tests (Fig. 4-5). These differences are not overcome by social class. Possibly the lower scoring is a result of decreased attention and verbal communication with individual children. In addition, there is an increase in the rate of mental retardation.

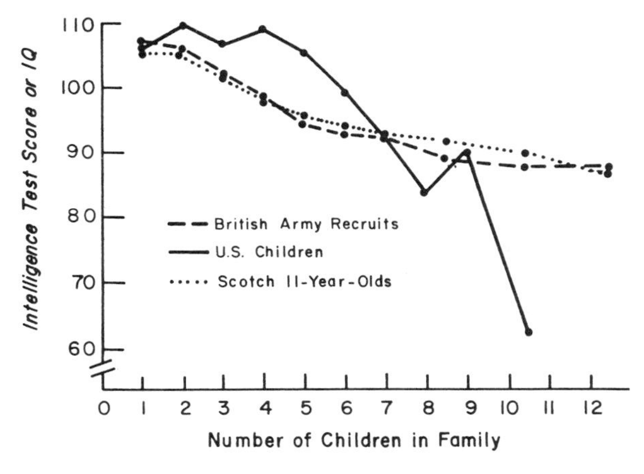

FIGURE 4-5. Variations of IQ or intelligence test scores of children in England, Scotland and the U.S. with relationship to family size. (Reprinted by permission of The Population Council from Wray, J. D.: *Population pressure on families: family size and child spacing.* Reports on Population/Family Planning. 9:427, 1971.)

For instance, there is a tenfold increase in the rate of mental retardation of the sixth child over the first child. Of interest is the fact that forty-seven percent of United States males rejected from military service come from families with six or more children. A report in the 1976 Social Security Bulletin in summary[2] says that children with fewer siblings are more likely to attend college, become professionals and have higher incomes while those from larger families are more likely to have less educational achievement, blue collar employment and lower yearly incomes. Moreover, young men and women from larger families tend to have larger families themselves.

All of these factors have strong implications for developing countries where approximately 25 percent of families have more than four children.

6. An increase in the rate of delinquency and other behavioral problems has been noted in children of large families.

Parental Health

Family size not only affects the health of child members of the family unit but also increases the mental and physical health problems of parents. In parents of large families, there is a definite increase in ulcer disease in males and in rheumatoid arthritis, diabetes mellitus and cancer of the cervix in females.

Advancing parity, regardless of age, presents a variety of obstetrical hazards to the mother. The occurrence of the following complications is increased in the grand multipara: death during pregnancy, labor and puerperium; placenta previa; uterine rupture; postpartum hemorrhage; abnormal presentations; and hypertensive disorders of pregnancy.

HEALTH CONSEQUENCES OF SHORT BIRTH INTERVALS

Consequences to Children

When there is a short interval between successive births of children in a family, there is a corresponding increase in fetal, neonatal, infant and childhood mortality. This is particularly applicable to a four-month's or less interval (Fig. 4-6). These values may vary depending upon the socioeconomic situation.

There is an increase in malnutrition on short birth interval babies. For instance, kwashiorkor, common throughout the subtropic world, is a severe nutritional disease of children and results from protein deprivation, as a result of weaning the child from the breast. The word, derived from African tribal language, means "the deposed baby when the next one is born."

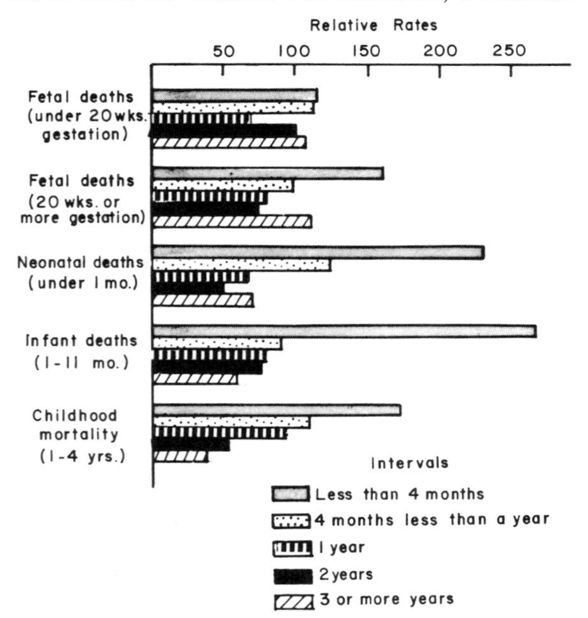

FIGURE 4-6. Occurrence of mortality in short birth intervals. (Adapted from Wray, J. D.: *Population pressure in families: family size and child spacing*. Report on Population/Family Planning 9:437, 1971.)

Maternal Consequences

Although there is inadequate evidence to substantiate a deleterious effect of short interval births on the mother, teleological considerations suggest a maternal depletion syndrome.

EXTREMES OF REPRODUCTIVE YEARS

In an attempt to find a solution to the problems of large family size, one should look at childbearing at the extremes of the reproductive years. By eliminating fertility of those under 18 and over 35 there would be a significant decrease in the world and United States birth rate (Figs. 4-7 and 4-8).

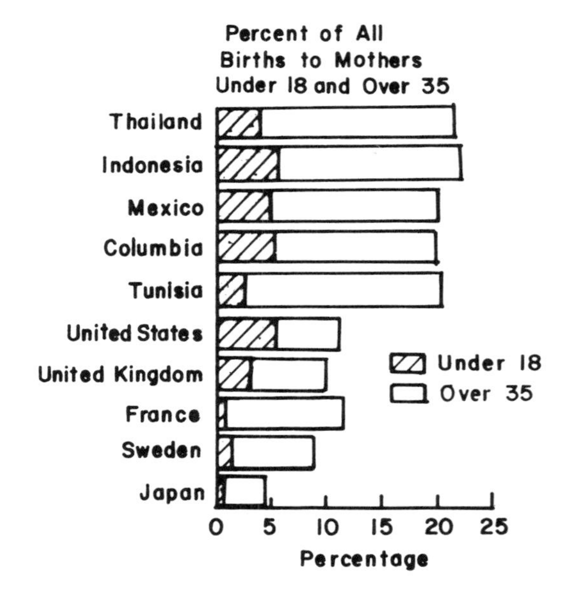

FIGURE 4-7. Percentage of all births to mothers under 17 and over 35 in various nations. (Adapted from Berelson, B.: Population Council Annual Report 1971.)

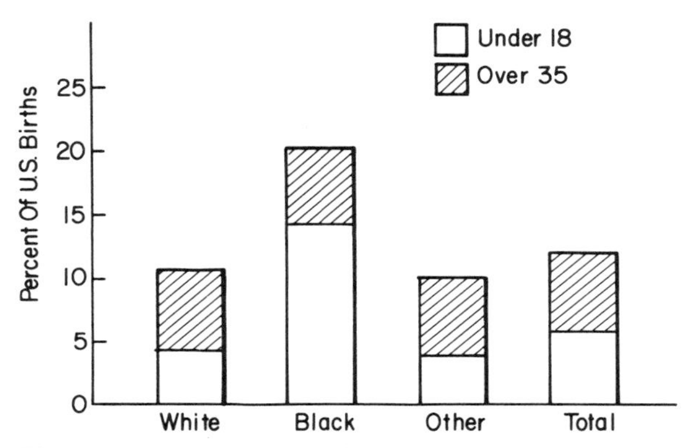

FIGURE 4-8. Percent of U.S. births to mothers under 18 and over 35. (Based on data from Berelson, B.: Population Council Annual Report 1971.)

Pregnancy Related Complications

Pregnancy related complications are significantly higher in these age groups.

1. Mothers conceiving between the ages of 30 and 40 have a twofold risk of pregnancy related death (Fig. 4-9). This increases to four- to fivefold if the female is over 40. For instance, in Thailand 40 percent of maternal deaths were in females over 35 although they accounted for only 20 percent of all births.

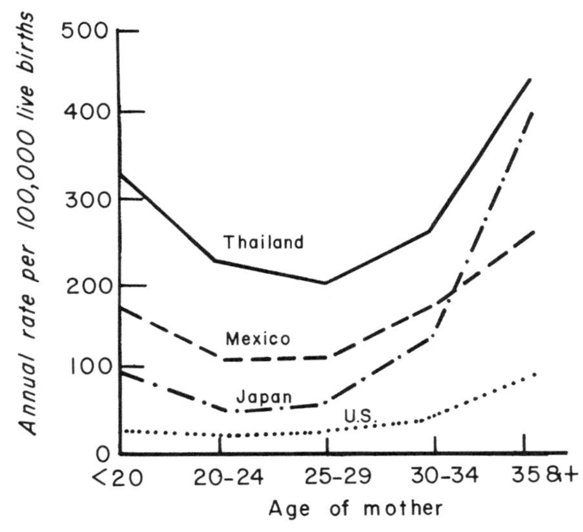

FIGURE 4-9. Maternal death rates for the period 1964 to 1966 for several nations. (Adapted from Berelson, B.: Population Council Annual Report 1971.)

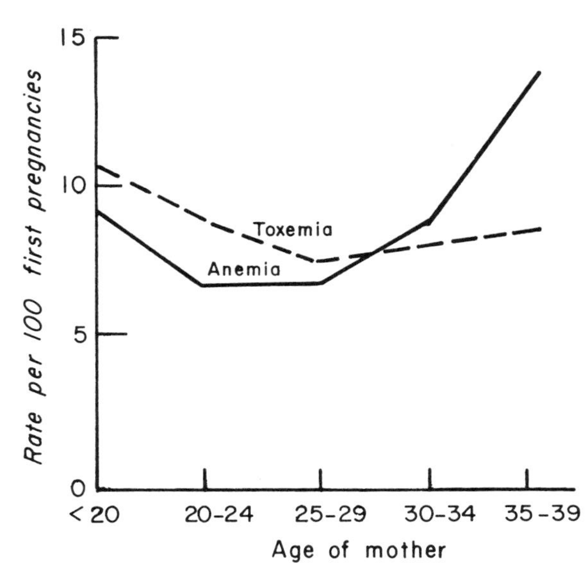

FIGURE 4-10. Complications of first pregnancy, U.S., 1961 to 1964 study. (Adapted from Berelson, B.: Population Council Annual Report 1971.)

2. Maternal morbidity is higher in these age group extremes. The decrease and increase in two major complications of pregnancy are shown in Figure 4-10.

3. Fetal and neonatal mortality (Figs. 4-11

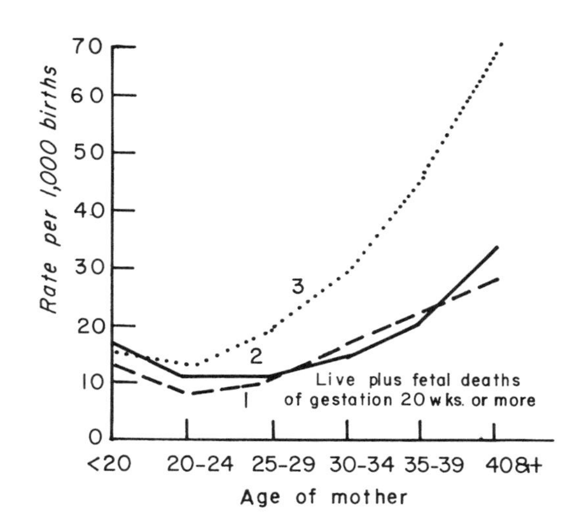

FIGURE 4-11. U.S. fetal death rates for birth orders 1, 2 and 3, 1963. (Adapted from Berelson, B.: Population Council Annual Report 1971.)

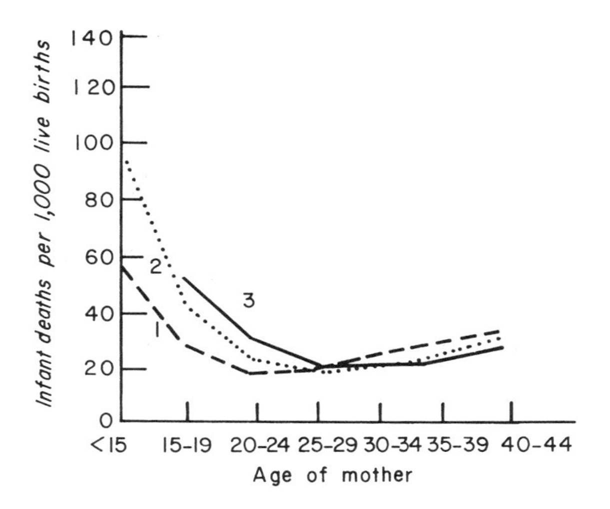

FIGURE 4-12. U.S. infant mortality, 1960 birth cohort for birth orders 1, 2 and 3. (Adapted from Berelson, B.: Population Council Annual Report 1971.)

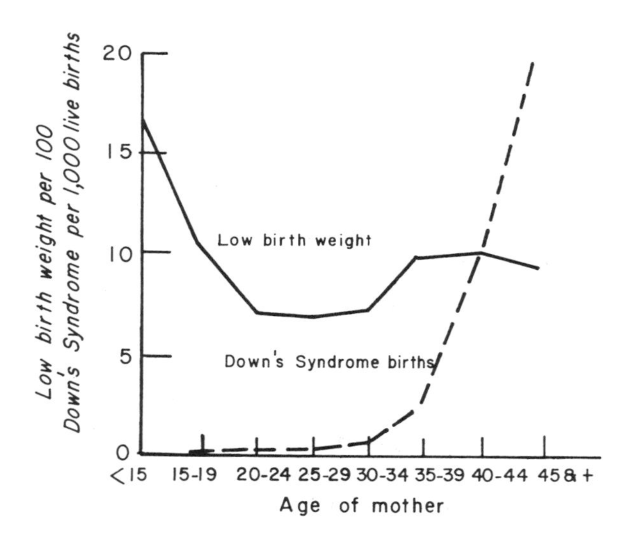

FIGURE 4-13. Low birth weight infants in the U.S. in 1967 and Down's syndrome births in Australia from 1942 to 1957. (Adapted from Berelson, B.: Population Council Annual Report 1971.)

and 4-12) and morbidity increase in women of these reproductive years. Notice in Figure 4-13 how low birth weight is related to the young age of the mother while Down's syndrome is related to an older maternal age.

FAMILY PLANNING CENTERS AS A SOLUTION

Many of the problems related to the three complications above have been shown to be ameliorated by specialized health care for women with this high risk potential. Elimination of the problem of decreased health care can best be achieved through the availability and utilization of family planning centers as primary health care units. The Social Security Amendment of 1972 requires that all states have available family planning services for Medicaid clients. Ninety percent of the expenditures by the states for family planning are reimbursed by the federal government.

Assessment of the short-term government expenditure (from 1966 to 1971), associated with those births to Medicaid patients that have been averted through family planning programs in the United States, indicate the following advantages and benefits:

1. medical care associated with the pregnancy
2. public assistance for children
3. opportunity costs of income lost to the mother (estimated at between $368 and $524 million dollars yearly). This showed an overall short-term benefit/cost ratio of between 1.8 : 1 and 2.5 : 1.

Note that this does not take into account all the economic, social and personal benefits to the individual and family. Nor does it take into account the long-term costs to society, which include education, welfare dependency, housing, health care and public services.

An extension of family planning services is also useful in families where genetic predispositions should be detected and future reproduction modified. Before assuming this responsibility, however, the health care provider should be aware of the complexities of the problem and should avail himself of the services of experienced genetic counselors.

Family planning facilities alone do not solve the problems connected with large families. Motivational factors that affect the desire of the individual to limit the number of children are paramount. Interestingly enough, this can be seen from the marked drop in fertility rates in the United States during the nineteenth and twentieth centuries *prior to the ease of availability and legalization of contraceptives and therapeutic abortions*. However, this should not detract from the influence of contraceptives and abortions.

SUMMARY

Excess reproduction places severe stresses on the welfare of the family. As family size increases, the services the family can give each child decreases. This can be seen in economic, social, educational and health factors. Short birth intervals and childbearing at extremes of the reproductive years add further complications. Family planning services are very important in the reduction of substantial health risks to mother and child, but motivational factors are of primary importance in fertility regulation.

QUESTIONS

1. What are the consequences of large family size on the individual child? Why?
2. What are the consequences of short intervals between successive births on the child? on the mother?
3. What hazards are associated with reproduction at the extremes of the reproductive ages?

REFERENCES

1. Omran, A. R.: *Health rationale for family planning.* In *Population Change, A Strategy for Physicians.* World Federation of Medical Education, Bethesda, Md., 1974, p. 75.
2. Orshanisky, M., and Bretz, N. J.: *Born to poor: birthplace and number of brothers and sisters as factors in adult poverty.* Social Security Bulletin 39:21, 1976.

BIBLIOGRAPHY

Berelson, B.: Report of the President, The Population Council Annual Report, 1971.

Dott, A. B., and Fort, A.: *The effect of maternal demographic factors on infant mortality rates.* Am. J. Obstet. Gynecol. 123:847, 1975.

Jaffe, F. S.: *Short-term costs and benefits of United States family planning programs.* Stud. Fam. Plann. 5:98, 1973.

Nortman, D.: Reports on Population/Family Planning. August 1974, p. 404.

Omran, A. R.: *Health benefits for mother and child.* World Health, January 1974.

Rickels, K., et al.: *Emotional assessment in obstetrics and gynecology.* In Young, D. (ed.): *Primary Care.* Williams and Wilkins, Baltimore. (In press.)

Wray, J. D.: *Population pressure of families: family size and child spacing.* Reports on Population/Family Planning. August 1971.

5 The Changing Status of Woman

Woman's role in society has been continuously changing. The most recent social changes have a significant demographic impact; conversely, demographic changes in society can significantly alter the status of women. These relationships will be explored in this chapter.

There are numerous examples historically which support the concept that women were relegated a lower status than that of men. Consider some of the following examples of customs which were observed by some nations, cultures or religions:

1. If the husband died, the wife was married to her husband's brother in order to continue the family name.
2. If the husband died, the wife was buried or cremated with his body.
3. The woman never appeared in public.
4. A man could have more than one wife.
5. Social custom scorns the sexually active female while glamorizing the male in similar activities.

Major cultural disparities persist among nations with regard to woman's role. Some of the customs were originally established to protect the woman or the family, while others were adopted in societies where women outnumbered men (e.g., early Mormons).

In the United States, a woman followed the traditional role of a woman observed by the country of her family origin. Eventually most adapted to the changing mores of the American culture. Her position in society still was derived mainly in terms of her relationship to men—whether she was wife, mother, sister or daughter.

In the United States, woman's right to vote has been guaranteed through an Amendment to the Constitution. This changed woman's status in the nation. The real beginning of the modern concept of womanhood came during World War I when factories and places of business were forced to hire women to replace men who were in the Armed Forces. When the war was over, many women decided to remain in the labor force, and customs in the United States continued to change. This did not insure that all the barriers to woman's emancipation were down. It was merely a beginning.

CHANGING BARRIERS

The interrelating barriers to social equality include procreation, lack of domestic alternatives, cultural labels, institutional definitions, social customs, lack of educational opportunities, occupational experiences and inadequate medical knowledge and facilities. These barriers are changing and with them the potential for changes in fertility.

In the following sections we present problems, situations and solutions for your study.

Procreation

"The status of women may be seen as both a determinant and a consequence of variations in reproductive behavior. A woman's health, educational opportunities, employment, political rights and role in marriage and the family may all affect and, in turn, be affected by the timing and number of her children and her knowledge of how to plan births."[1]

Birth expectations of American women are decreasing. For instance, 58 percent of married women ages 18 to 39 desire two or fewer children during their lifetime, and total birth expectation of wives 18 to 39 has decreased from 3.1 in 1967 to 2.2 in 1974.[2]

Domestic Alternatives

There has been a lack of alternatives for women outside of the home. The subject of occupational experiences is discussed later in this chapter. For a long time there was hardly anything else for a woman to do but stay at home—or work in someone else's home as a domestic.

The sharing of domestic responsibilities in the home increases in direct proportion to the woman's educational and employment status.[3]

Cultural Labels

The traditional definition of "woman's role" as a mother and homemaker persists within our society, strongly limiting a woman's choices. Special attention has been given recently to the traditional role expectations perpetrated in children's books and commercial advertisements.

This role definition is supported by both sexes. Frequently it is pointed out that the lack of responsibility and understanding of the male in fertility control leads to a continuance of the woman in her domestic role. But it also must be noted that women, unfortunately, often support such practices.

Institutional Definitions

Labels continue to be placed on institutions or by institutions. The changing status of these institutions or of these labels should be studied too.

Marriage. Universal marriage in certain parts of Asia and Africa is associated with high levels of fertility in those countries. The decline in birth rates in Western Europe during the late nineteenth and early twentieth century was associated with late marriage and the high proportion of unmarried women (see Chapter 2). Presently the number of women in the United States electing not to marry is increasing as is also the ease in separation and divorce (see Chapter 3). This undoubtedly could affect the fertility in the United States but would be influenced by the specific social situations.

The Family. The concept and makeup of the family unit is also changing. One of the indications of this is the acceptance in adoption cases of single parents. Another is the number of women who are leaving the family, where it previously has been the man who walks out.

The Church. All religions have long fostered practices which have led to an inequality among the sexes. Conservative religions express restrictive stands on abortion and birth control and many religious groups forbid the ordination of women as ministers.

Legal Institutions. Laws in many countries foster inequalities among the sexes. Recent efforts in the United States to enact an Equal Rights Amendment have been met with considerable unexpected opposition.

Government and Politics. It is only in recent decades that women have been able to vote or run for office. Despite the large percentage of female voters, the number of female legislators is miniscule. In 1975, there were *no* female United States Senators, only 1 Governor and a mere 17 federal Congresswomen.

Social Customs

There are social barriers to change which in themselves create inequities. Again it should be noted, these inequities are mutually perpetrated by both sexes. Consider how the following have changed or remain as barriers: dating customs, social amenities and mass media.

Educational Opportunities

Educational opportunities for women have strong bearing on subsequent fertility and economic achievement. The following examples should be considered:

1. Next to age, the educational level of women appears to be one of the strongest factors affecting fertility. The fertility of a particular couple correlates best with the wife's education. Also, education opens a greater range of options and interests outside the home.[4]
2. In the United States, marriage and childbearing have been known to have a greater negative effect on women's education than on men's.
3. Illiteracy rates in most countries are much higher among women than among men (see Fig. 1-2).
4. In most countries females are less than half of the school population with even lower proportions at higher levels of training.

Occupational Experiences

Work opportunities for females have great influence on subsequent reproductive behavior just as fertility influences occupational achievements. These ideas can be observed in Table 5-1, which shows the percentage of economically active females in various countries.

Delay of marriage and childbearing may allow a woman to pursue vocational skills, thus enabling her to obtain skills necessary for better employment. Women employed full time or who work for a major part of their lives have fewer children. The effect on fertility is significant. Working women desire, and actually have, fewer children than nonworking women.[5] Also, women who work in more economically and personally gratifying jobs (e.g. professional, white collar) have smaller families than women who work in service-related or blue collar jobs.[6] Major declines in the birth rate have often preceded an increase in the participation of women in the labor force.

There is an increase in the nonfarm work force of women in the United States (Fig. 5-1).[7] The percentage of married women in the United States who work is growing (Fig. 5-2). This increase in the number of working females (the labor force) has been influenced to a great extent by education, inflationary pressures, smaller families, increased divorce rate and women's liberation movement. It is interesting to observe the numerous determinants which send women into the labor force.

Despite this increase in the number of females in the United States labor force, it should be noted that (1) women have typically participated in economic activity which has

Table 5-1. Percentage of economically active females.

Country	*Economically active females*	*Proportion to male economic activity rates*
Africa	26%	53%
Asia	22%	43%
Americas*	18%	36%
Oceania	16%	31%
Soviet Union	45%	86%
Europe	30%	51%
Muslim countries	5%	7%
United States	30%	56%

Data from Dixon, R.B.: *Women's rights and fertility*. Reports on Population/Family Planning. January 1975, p. 2.
*Including Canada and United States

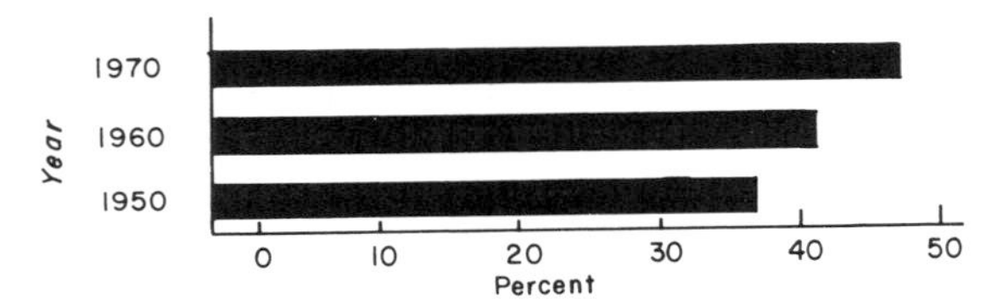

FIGURE 5-1. Change in the U. S. female nonfarm work force, 1950 to 1970. (Adapted from Blake, J.: *The changing status of women in developed countries*. Sci. Am. 231:138, 1974.)

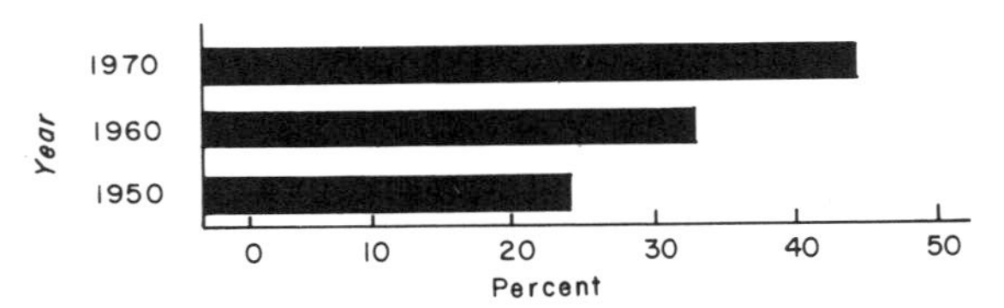

FIGURE 5-2. Percentage of married women who work. (Adapted from Blake, J.: *The changing status of women in developed countries*. Sci. Am. 231: 142, 1974.)

supplemented their primary status within the home and (2) there is a differential in the work experience. In the United States during 1973, nearly half of the female labor force worked fewer than 35 hours per week. The differential in the work experience is shown in Figure 5-3. Study each category for both sexes. The high proportion of females in professional and technical fields is misleading since most of these professions include such occupations as teaching, nursing and social work. The proportion of females to males in medicine is 7 percent and in dentistry 1 percent.

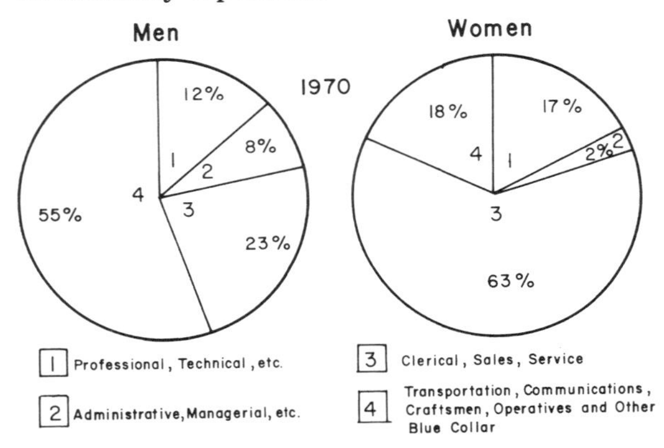

FIGURE 5-3. Skewed occupational distribution with sex-typing of jobs. (Adapted from Blake, J.: *The changing status of women in developed countries*. Sci. Am. 231:145, 1974.)

However there has been a change in the percentages of females in certain professions from 1960 to 1970. There has been a 62 percent increase (from 7 to 9 percent) of female physicians and a 149 percent increase (from 2.4 to 4.7 percent) of women as lawyers. In 1974, 18 percent of medical students and 19 percent of law school students in the United States were female.[8]

Social problems encountered by women in the labor force include

1. inequity of pay
2. lack of facilities for child care
3. increased work load (work plus housework plus childcare) and decreased leisure
4. expanded demand for women's services dependent upon high rates of economic growth (In periods of economic recession, women are among the first to lose their jobs)

Medical Knowledge and Facilities

Both opponents and proponents of woman's emancipation make themselves known through medical knowledge and facilities. The opponents burden females with unwanted pregnancies and frustrated lives. The proponents provide good medical facilities, the knowledge and opportunities to plan pregnancies and postpone fertility, and an opportunity for the female to seek her goal as a person.

If inadequate facilities for provision of family planning are provided, females are hampered in their desire to control their fertility. Legislators of both sexes have imposed restrictive legislation, which includes anti-abortion laws and restriction on sex and family planning education in the schools. Special interest, well-financed religious groups further limit a woman's ability to effectively regulate her fertility.

Proponents of woman's equality recognize the right of all women to control their own bodies and reproductive destinies through access to high-quality, comprehensive medical care and health education.

Postponement of fertility potentially decreases the opportunity for fertility. (This will be discussed in Chapter 7.) The biologic results of postponement are immediate (women at younger ages are more fertile) and long range (decreasing fertility with increasing age). Postponement allows the female to improve her education and to develop interests apart from home and family.

SEXUAL ASSAULT

The health care worker who deals with aspects of human sexuality should be familiar with the problems of sexual abuse in the female where reproductive implications must be carefully considered. The person who cares for the rape victim should have both a systematic approach to the care of the victim for legal requirements and a sensitivity to the medical and psychological needs of the patient.

Rape is defined as a form of sexual assault which includes labial penetration by the penis without consent. Ejaculation need not occur.

Rape is the most frequently committed violent crime, exceeding the sum total of aggrevated assaults and homicides. Although nearly 50,000 rapes are reported each year in the United States, it is estimated only one rape in ten is actually reported. Approximately one woman in 200 will be raped this year in the United States.[9]

Of all major crimes committed in the United States, rape has the lowest conviction rate. The chances of apprehension are slim and, if the rapist is caught, the chances of conviction are even less. Furthermore, in the courtroom, the victim rather than the assailant is often on trial. For instance, in the city of Philadelphia for the year 1972 there were 860 complaints, 400 apprehensions, 200 trials but only 22 convictions. Assuming that only one in ten rapes are reported, the assailant has a 0.25 percent chance of conviction.[10]

Listed below is some factual information about rape. Statistics on 800 rapes were compiled by Women Organized Against Rape (WOAR) for the city of Philadelphia for the year 1974.

1. Fifty-four percent of rapes involved the use of a weapon, generally a knife.
2. Fifty-two percent of rape victims had visible signs of physical violence. In 12 percent of the victims the trauma was severe enough to require medical care.
3. Twenty-five percent of rape victims were adolescents, ages 13 to 16; this was the largest single group.
4. Fifty-seven percent of rape victims know their assailant; 27 percent well and 38 percent superficially. Eight percent of rapes are by family members. Forty one percent of the latter victims are in the 16-and-under age bracket.
5. Thirty-six percent of rapes occur on Friday and Saturday; 63 percent occur after dark.
6. Thirty-two percent of rapes are gang rapes; 90 percent of these are premeditated.
7. Forty-nine percent of rapes occur in a home—31 percent in the victim's home and 18 percent in the assailant's home. Twenty percent occur outdoors, 5 percent in cars and 26 percent in various indoor areas.
8. Twenty-seven percent of rape victims were white, 73 percent black. In 80 percent of rapes both the assailant and the victim are of the same race. In 17 percent of cases the assailant is white and the victim black. In 3 percent of the cases the assailant is black and the victim white.

Care of the Rape Victim

Immediate Care of Physical Injuries. Rape victims may experience a variety of traumatic injuries, both physical and emotional. The face and vagina are the most common sites of physical injury. All such injuries should be x-rayed and treated.

Medicolegal Documentation. This documentation includes the following:

1. A history of the assault should be obtained by the physician. Even though it is not used as evidence in most cases, it is useful in treatment of the victim.
2. In order to assess the risk of pregnancy the physician should obtain and record a menstrual, contraceptive, pregnancy and sexual history. A pregnancy test should be performed if

there is a suggestion of possible early gestation.

3. The physician must then examine all body surfaces and describe in detail any evidence of trauma. Allegations of anal or oral contact will necessitate careful search of these areas. This report may be the only evidence the patient has to support her claims of trauma.

4. The physician should also carefully describe the emotional and mental state of the patient. Blood and urine specimens for alcohol or drugs should be obtained if such are suspected.

5. A complete record of the pelvic examination should be made. Vulvar, vaginal and cervical smears should be made. The vagina should be rinsed with saline and the collection separated and studied for acid phosphatase. Specimens should be sent to a laboratory which is experienced in handling these specific studies. Where possible the cervical mucus should be examined under the microscope for sperm. Sedation may be required for examination of the pediatric patient.

Venereal Disease Prophylaxis. The following steps should be taken for prophylaxis:

1. A baseline serum serology is drawn.
2. Cervical and urethral cultures are obtained and placed on Transgrow or Thayer-Martin media and incubated in an anaerobic atmosphere.
3. The patient is given 1.0 gm. Probenamid and 4.8 million units procaine penicillin intramuscularly. (If there is a history of penicillin allergy, tetracycline, erythromycin or spectinomycin may be substituted.
4. On follow-up examination in four weeks a repeat serology and cultures should be obtained.

Pregnancy Prevention. If, because of the attack, there is a possibility that pregnancy exists, postcoital contraception should be offered. The greatest experience exists with the use of diethylstilbestrol 25 mg. orally twice daily for five days beginning immediately. (See section on postcoital contraception in Chapter 7.)

The patient is examined at a follow-up visit and, if necessary, tested to rule out pregnancy.

Prevention of Psychological Damage. A careful explanation of all examinations and the reasons for them should be given to the patient. The acute psychological effects of the rape may not be apparent to the examiner. One should not assume that a quiet unruffled female is not indeed experiencing psychological trauma. The victim needs reassurance not only in dealing with the immediate circumstances but also in her subsequent meeting and living with her family, friends or sexual partner. The physician should provide this support. Mild sedatives will be helpful where necessary.

In Philadelphia a volunteer group, Women Organized Against Rape (WOAR), has a female volunteer on call 24 hours a day to help the rape victim. Volunteers provide the victim with emotional support, legal advice and companionship during the post-assault ordeal. They also help the victim if requested during the subsequent trial. Rape crisis centers exist in many other cities and their assistance should be sought.

Follow-up can also be made on request one to two weeks after the assault by psychiatric social workers. Appropriate referral is made if necessary.

Long term psychological problems may result from the attack. The victim may suffer from acute or chronic anxiety or paranoia following the attack and she may develop adverse ideas about sex and her sexuality. The physician should be aware of the possibility of these problems.

SUMMARY

The ability of each woman to control her own fertility has been sharply limited by a variety of social barriers. These barriers include procreation itself, lack of domestic alternatives, cultural labels, institutional definitions, social customs, lack of educational opportunities, occupational experiences, and inadequate medical knowledge and facilities. Fertility itself has served as a barrier to social equality for many women. Woman's role in society is changing. With these transitions may come a change in fertility, which may have a profound effect on the world population.

Rape is the most frequently committed violent crime. Anyone who deals with aspects of human sexuality should be familiar with the medical, legal and psychological care of the rape victim. Rape crisis centers exist in many cities; their assistance should be sought by the health care provider.

QUESTIONS

1. What were the pressures that encouraged a woman to reproduce?
2. What are the advantages and disadvantages of reproduction for the woman?
3. What social changes should be enacted to enable females the right to determine their own fertility?

REFERENCES

1. Dixon, R. B.: *Women's rights and fertility*. Reports on Pop. Fam. Plann. January 1975, p. 2.
2. *Prospects for American Fertility*. U. S. Bureau of Census. June 1974.
3. Dixon, R. B.: op. cit.
4. Ibid.
5. Pratt, C., and Whelpton, P. K. (eds.): *Social and Psychological Factors Affecting Fertility,* Vol. 5. Milbank Memorial Fund, New York, 1958.
6. Dixon, R. B.: op. cit.
7. Blake, J.: *The changing status of women in developed countries*. Sci. Am., September 1974.
8. *The Philadelphia Inquirer*. July 11, 1975, p. 1.
9. Statistics compiled by Women Organized Against Rape (WOAR), Philadelphia, 1974.
10. Ibid.

BIBLIOGRAPHY

Massey, J. B., Garcia, C. R., and Emich, J. P.: *Management of sexually assaulted females*. Obstet. Gynecol. 38:29, 1971.

Rosenfeld, D. L., and Garcia, C. R.: *Medical aspects of rape*. Human Sexuality. March 1976, p. 77.

6 Adolescent Sexuality

The care of the adolescent patient presents special problems to the physician. Most strikingly there is a difficulty in communications. The adolescent female tends to be more impetuous, anxious and far less sophisticated than the older woman. Frequently her reason for coming is to have her initial examination or she has a "crisis related" problem. Her anxieties are compounded by societal and/or familial pressures as well as an anticipated moral judgment from the health care provider. Therefore it is not surprising that adolescents have the lowest rate of physician visits each year.

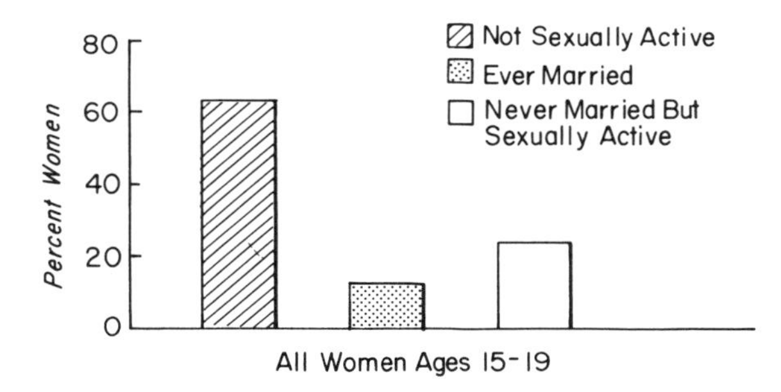

FIGURE 6-1. Relative proportion of females aged 15 to 19 who are sexually active, U.S., 1970. (Based on data from Whelan, E. M., and Higgins, G. K.: *Teenage Childbearing: Extent and Consequence*. Consortium on Early Childbearing and Childrearing, Child Welfare League of America, Inc., January 1973.)

SEXUAL AND SOCIAL PATTERNS

There are approximately 2 to 4 million never-married females ages 15 to 19 in the

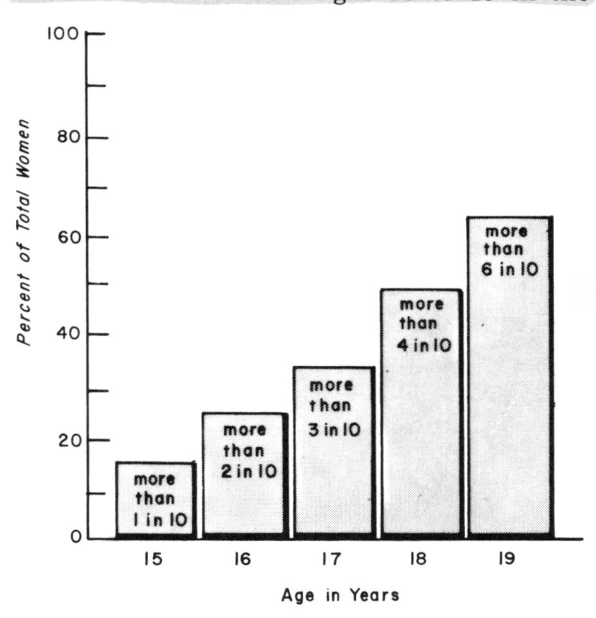

FIGURE 6-2. Relative proportion of females aged 15 to 19 who are sexually active by specific age, U.S., 1970. (Adapted from Whelan, E. M., and Higgins, G. K.: *Teenage Childbearing: Extent and Consequence*. Consortium on Early Childbearing and Childrearing, Child Welfare League of America, Inc., January 1973.)

United States; of these about 28 percent are sexually active. The sexually-active percentage rises to 37 if one is to include married or ever-married women in that age group (Figs. 6-1 and 6-2). One third of the women in this age group have had at least one pregnancy outside of marriage. In fact, nearly a third of the legal abortions in the country are performed on women under 20 (Fig. 6-3).

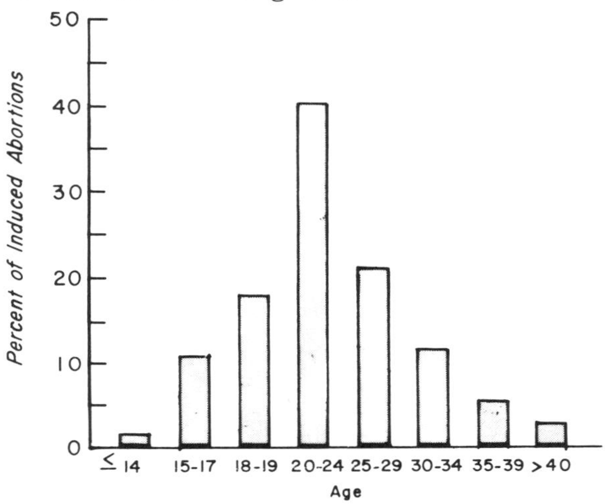

FIGURE 6-3. Age distribution of patients with induced abortion. (Adapted from Tietze, C., and Lewit, S.: *Early complications of abortions under medical auspices: a preliminary report.* Stud. Fam. Plann. 2:138, 1971.)

As a result of the baby boom (1946 to 1957) the number of teenagers in the United States is rising both in numbers (Figs. 6-4) and in proportion to the total population. (Note how the number almost doubled in a period of 20 years.) This rise is expected to taper off, however, by the end of the decade.

Those who provide contraceptive services to adolescents should be familiar with their sexual habits and the social variables that have bearing on these habits. The first part of this chapter deals with coital patterns, racial and socioeconomic patterns and contraception.

Coital Patterns

Recent reviews on surveys on adolescent sexual behavior and attitudes[1,2,3] have provided a better insight into the coital patterns of this age group. Various health care providers

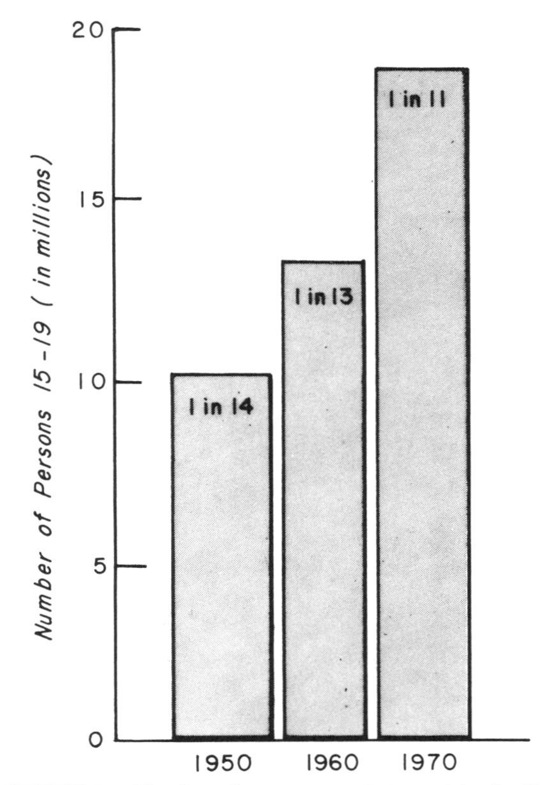

FIGURE 6-4. Number of persons aged 15 to 19 in the U.S., 1950, 1960, 1970. (Adapted from Whelan, E.M., and Higgins, G. K.: *Teenage Childbearing: Extent and Consequence.* Consortium on Early Childbearing and Childrearing, Child Welfare League of America, Inc., January 1973.)

would benefit from familiarity with this information summarized in the following:

1. Sexually-experienced adolescents (both male and female) tend to be in stable relationships at this time in life.
2. Sixty percent of adolescents have had intercourse with only one partner. For whites, 50 percent of 19-year olds and 70 percent of 15-year olds have had only one partner. For blacks, 56 percent of 19-year olds and 65 percent of 15-year olds have had only one partner.
3. Fifteen percent of whites and 11 percent of black sexually-active adolescents have had more than three partners.
4. Approximately 50 percent of sexually-active adolescent females interviewed indicated they only had relations with the man they intended to marry.

Sexually-active adolescent females experience coitus with modest frequency. For comparison, the coital frequency of married women less than 25 years of age is nine times or more per month, while the coital frequency of sexually-active adolescents for the month prior to interview is not as frequent (Fig. 6-5). This coital activity for unmarrieds increases with age (Fig. 6-6).

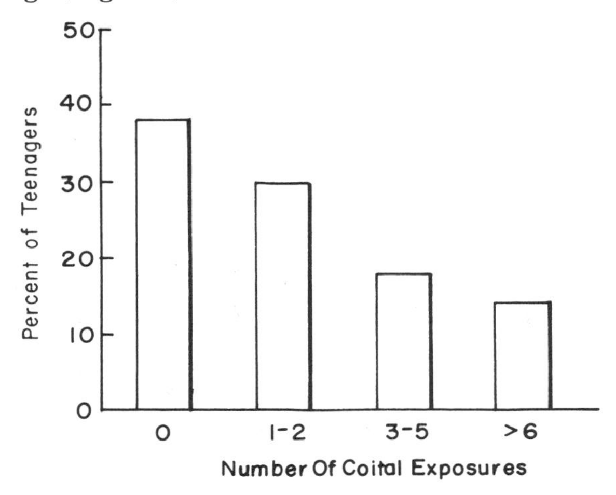

FIGURE 6-5. Monthly coital frequency of sexually-active adolescent females. (Based on data from Kantner, J. F., and Zelnick, M.: *Sexual experience of young unmarried women in the United States*. Fam. Plann. Perspect. 4:1, 1972.)

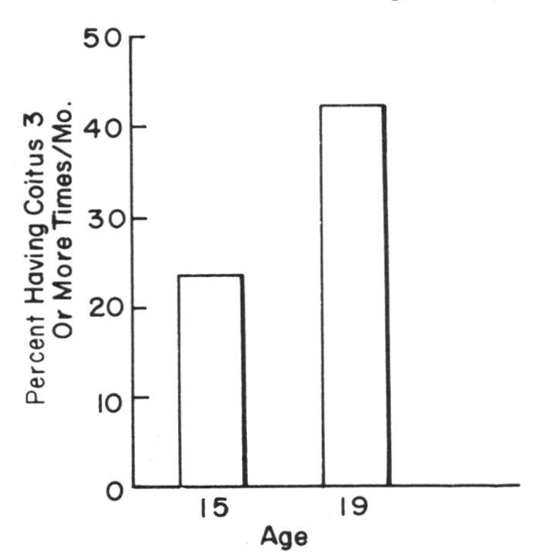

FIGURE 6-6. Percentage of adolescents having coitus three or more times per month. (Based on data from Kantner, J. F., and Zelnick, M.: *Sexual experience of young unmarried women in the United States*. Fam. Plann. Perspect. 4:1, 1972.)

Social Patterns

The chief racial differences in coital patterns are in the proportions who had sexual experience and the age at which coitus began. These proportions are twice as high for blacks at each age level. Among sexually-active adolescents, however, blacks have intercourse somewhat less frequently and with fewer partners than whites.

Coital parameters show an inverse relationship with socioeconomic background, level of family education and family income. The racial differences, however, remain within each category.

The lowest proportion of teenagers with coital experiences is found among females living in families headed by natural fathers. The highest proportions are among those teenagers living alone or apart from the family.

Coital exposure is less frequent in females who have open communication with their families and in those who are active in church activities. Highest coital activity is found in adolescents from a migrant background and those living within the inner city.

Fertility Control Among Adolescents

Frequency of Utilization. Remembering that approximately a third of sexually-active adolescent females have had at least one pregnancy outside of marriage, one should not be surprised that 53 percent of sexually-active adolescents failed to use any form of contraception the last time they had coitus. As the female becomes older (perhaps more experienced and/or better educated), the use of contraceptives increases (Figs. 6-7 and 6-8).

Factors Which Influence Utilization. Better contraceptive utilization correlates favorably with

1. higher socioeconomic background
2. better educational status of parent or guardian
3. household headed by natural father (This includes 63 percent of white and 35 percent of black sexually-experienced teenagers.)
4. higher family income

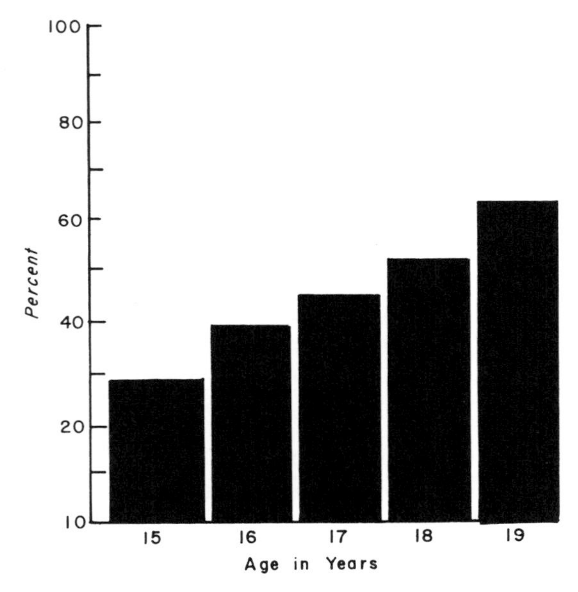

FIGURE 6-7. Proportion of never-married females aged 15 to 19, reporting the use of contraceptives at last intercourse, U.S., 1971. (Adapted from Whelan, E., and Higgins, G. K.: *Teenage Childbearing: Extent and Consequence*. Consortium on Early Childbearing and Childrearing, Child Welfare League of America, Inc., January, 1973.)

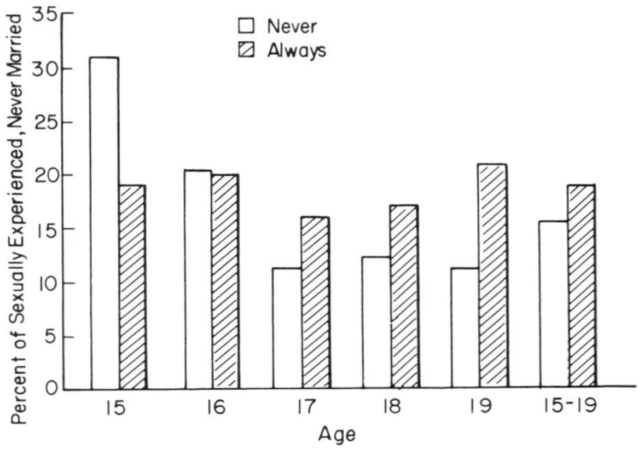

Never
Always
Percent of Sexually Experienced, Never Married
35
30
25
20
15
10
5
0
15
16
17
18
19
15-19
Age

FIGURE 6-8. Percent of sexually-experienced never-married females aged 15 to 19, according to contraceptive use shown by age. (Based on data from Kantner, J. F., and Zelnick, M.: *Contraception and pregnancy: experience of young unmarried women in the United States*. Fam. Plann. Perspect. 5:11, 1973.)

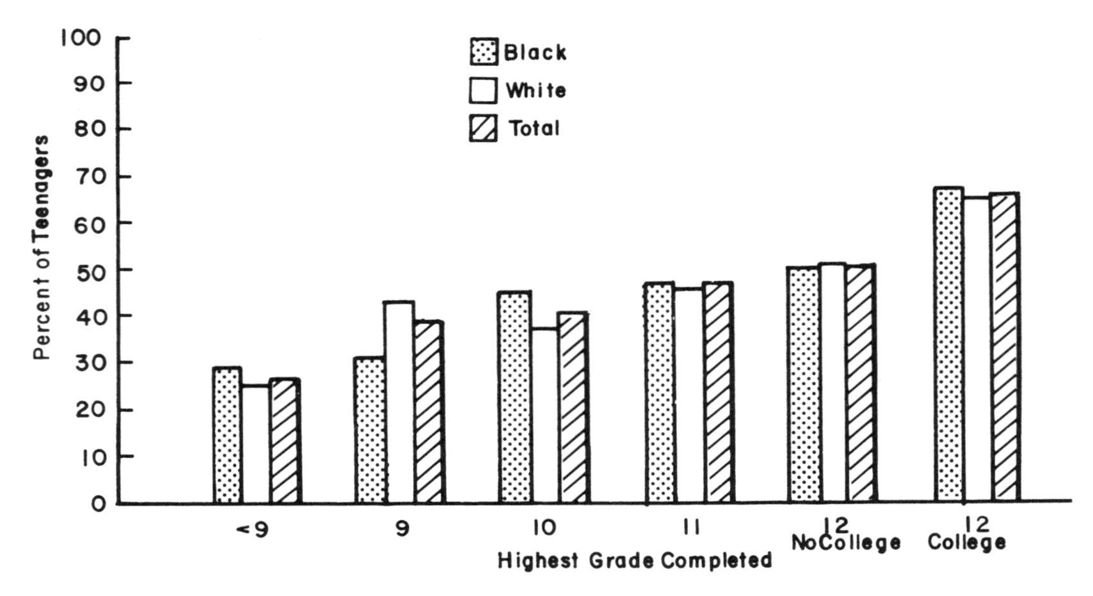

FIGURE 6-9. Contraception used at last intercourse with respect to race and education for sexually-active never-married adolescents. (Based on data from Kantner, J. F., and Zelnick, M.: *Contraception and pregnancy: experience of young unmarried women in the United States*. Fam. Plann. Perspect. 5:20, 1973.)

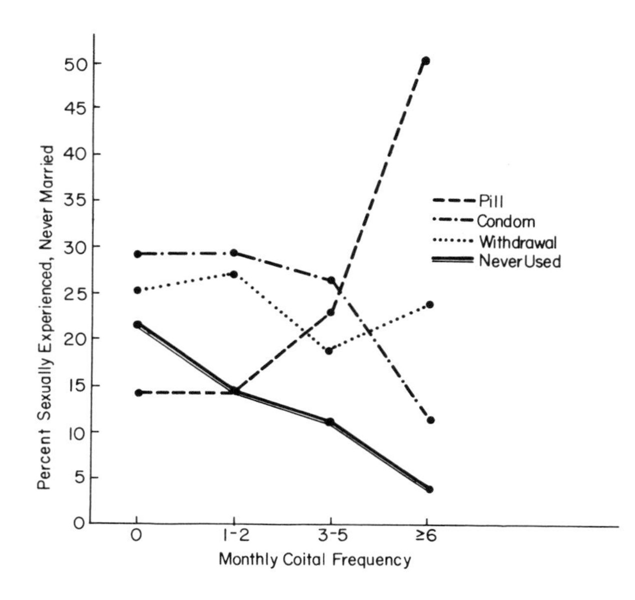

FIGURE 6-10. Percent of all sexually-experienced never-married females aged 15 to 19 according to contraceptive method most recently used. (Based on data from Kantner, J. F., and Zelnick, M.: *Contraception and pregnancy: experience of young unmarried women in the United States*. Fam. Plann. Perspect. 5:20, 1973.)

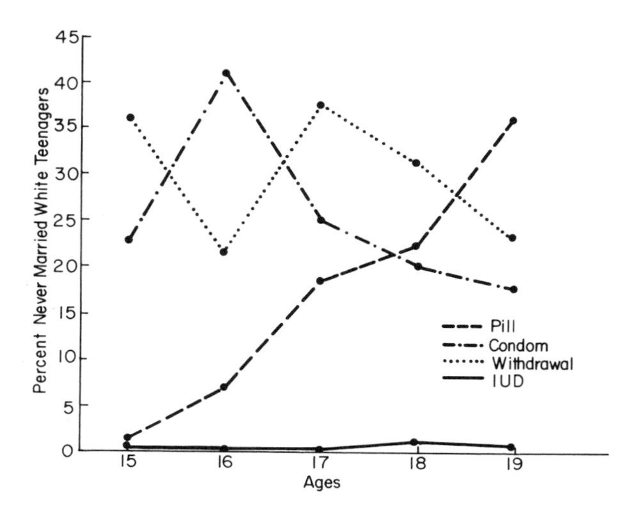

FIGURE 6-11. Percent of all white sexually-experienced never-married females aged 15 to 19 according to method most recently used. (Based on data from Kantner, J. F., and Zelnick, M.: *Contraception and pregnancy: experience of young unmarried women in the United States*. Fam. Plann. Perspect. 5:20, 1973.)

5. level of education of adolescent (Fig. 6-9)
6. age of adolescent
7. coital frequency

Poor contraceptive utilization correlates positively with the episodic nature of coital exposure. The 18- and 19-year olds whose coital frequency parallels that reported for married couples (greater than five times per month) utilize contraception at rates (72 and 78 percent respectively) which are similar to those for married couples. However, fewer than 14 percent of this group reported having coitus more than five times monthly.

Patterns of Fertility Control. *Method Chosen.* Study Figures 6-10, 6-11, and 6-12. From these illustrations we can make the following statements:

1. Oral contraception is used more frequently with increasing age.
2. Black teenagers use condoms and douching more than white teenagers.

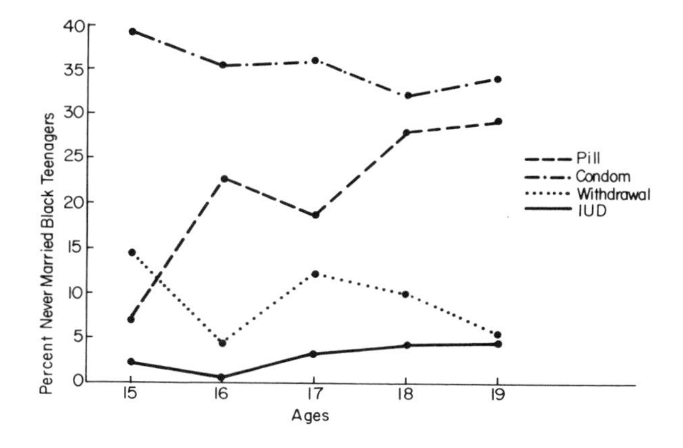

FIGURE 6-12. Percent of all black sexually-experienced never-married females aged 15 to 19 according to method most recently used. (Based on data from Kantner, J. F., and Zelnick, M.: *Contraception and pregnancy: experience of young unmarried women in the United States*. Fam. Plann. Perspect. 5:20, 1973.)

3. Withdrawal is far more common among white teenagers.

We can further state that married teenagers tend to utilize more effective means of contraception than unmarried teenagers (Fig. 6-13).

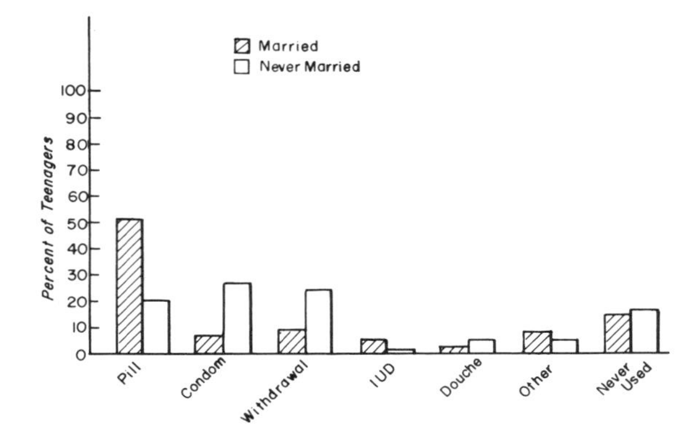

FIGURE 6-13. Contraceptive method most recently used by all sexually-active teenagers, whether married or never married. (Based on data from Kantner, J. F., and Zelnick, M.: *Contraception and pregnancy: experience of young unmarried women in the United States*. Fam. Plann. Perspect. 5:24, 1973.)

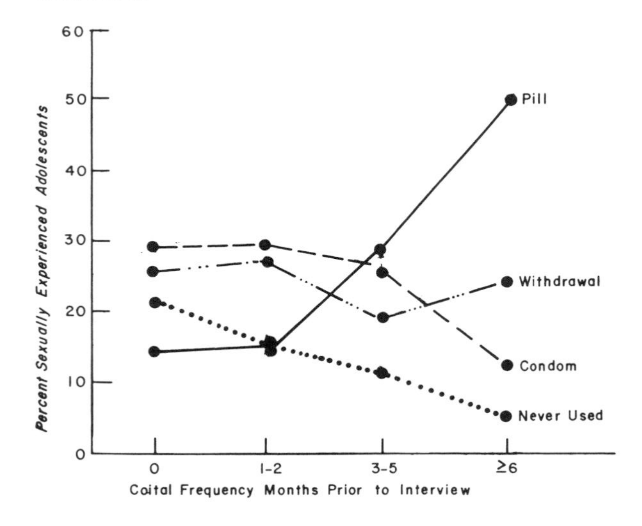

FIGURE 6-14. Percent of sexually experienced never-married females aged 15 to 19, according to method most recently used by monthly frequency of intercourse. (Based on data from Kantner, J. F., and Zelnick, M.: *Contraception and pregnancy: experience of young unmarried women in the United States*. Fam. Plann. Perspect. 5:20, 1973.)

Factors Which Influence Method of Contraception. There are no major statements that can be made concerning the reason for selection of certain contraceptives over others. However, two factors stand out: (1) there is no good correlation with socioeconomic status and type of contraceptive chosen, and (2) use of birth control pills tends to increase with an increase in coital frequency (Fig. 6-14).

Efficacy of Choice. Choice of a medically administered contraceptive (pill, IUD, diaphragm) increases by a factor of four the utilization of the method. For further illustration, study Table 6-1.

Table 6-1. Contraceptive utilization among teenagers.

Method used at last intercourse	*Ratio of users to non-users*
Pill, IUD, diaphragm	3.7
Condoms, foam	1.1
Withdrawal	0.6
Douche, rhythm	0.4

Adapted from Kantner, J. F., and Zelnick, M.: *Contraception and pregnancy: experience of young unmarried women in the United States*. Fam. Plann. Perspect. 5:24, 1973.

Sources for Contraception. For rather obvious reasons, married adolescents rely more than non-married adolescents on physicians for their methods of contraception (Fig. 6-15).

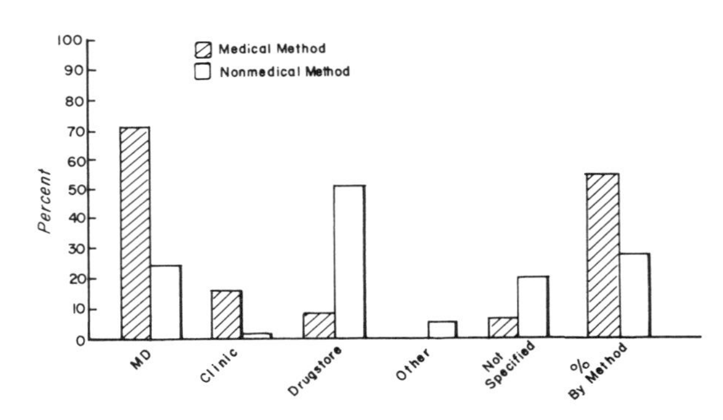

FIGURE 6-15. Percent of ever-contracepting ever-married females 15 to 19, according to source of contraception by type of method most recently used. (Based on data from Kantner, J. F., and Zelnick, M.: *Contraception and pregnancy: experience of young unmarried women in the United States*. Fam. Plann. Perspect. 5:24, 1973.)

The unmarried female relies on the drugstore; herein lies the difficulty the non-married adolescent female faces—reliance on less effective, non-medical methods (Fig. 6-16). Nonavailability is second to the most common reason for contraceptive non-utilization among unmarried teenagers, which is the false trust that pregnancy will not occur during that exposure.

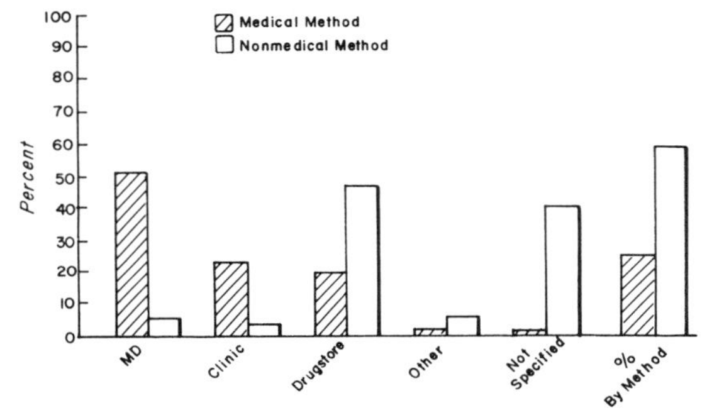

FIGURE 6-16. Percent of ever-contracepting never-married females aged 15 to 19 according to source of contraception by type of method most frequently used. (Based on data from Kantner, J. F., and Zelnick, M.: *Contraception and pregnancy: experience of young unmarried women in the United States.* Fam. Plann. Perspect. 5:21, 1973.)

BASIC KNOWLEDGE OF REPRODUCTION AND CONCEPTION

The following statements and facts are testimony to the ineffectiveness of the educational system in the United States as well as sex education in the home.

Among 15- to 19-year old never-married females, the major reasons for not using contraception[4] were

1. conception could not occur because sex was too infrequent
2. wrong time of the month for conception to occur
3. the girl was too young to become pregnant
4. the male claimed sterility

Study Table 6-2 for percentages of adolescents giving six reasons for not using contraceptives.

There is a direct relationship of the basic understanding of reproductive function of the adolescent with the level of education of the parent (Fig. 6-17). Nevertheless, the basic understanding of the menstrual cycle is poor. For instance, 33 percent of 19-year old white females whose mothers have college educations cannot identify the period of greatest risk. And 60 percent of all never-married teenagers believe that the time a girl can become pregnant is as soon as her period begins.

Twenty-eight percent of white and 55 percent of black sexually-experienced young women report that they do not believe they can become pregnant "easily." Of this group, 70 percent did not use contraception at last exposure. However, of those most strongly convinced that they do become pregnant easily, 68 percent used contraception at last intercourse.

Table 6-2. Percent of never-married sexually-experienced women 15 to 19 years of age according to reasons they reported for not using contraception (1971).

Reason	*Percent*
Time of month	39.7
Low risk	30.9
Nonavailability	30.5
Hedonistic objection	23.7
Desired pregnancy	15.8
Moral/medical objection	12.5

Adapted from Shah, F., Zelnick, M., and Kantner, J. F.: *Unprotected intercourse among unwed teenagers.* Fam. Plann. Perspect. 7:39, 1975.

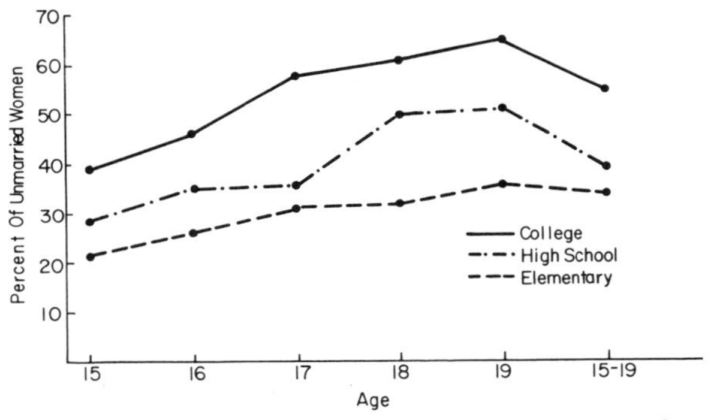

FIGURE 6-17. Percent of unmarried young females aged 15 to 19 who correctly perceive the time of greatest risk within the menstrual cycle for single years of age and education of female parent or guardian. (Based on data from Kantner, J. D., and Zelnick, M.: *Sexual experience of young unmarried women in the United States.* Fam. Plann. Perspect. 4:9, 1972.)

Table 6-3. Percent of live births and out-of-wedlock births to adolescents in various countries.

Country	*Year*	*Percent of live births*	*Percent of out-of-wedlock births*
Austria	1968	12.3	34.3
Bulgaria	1968	18.2	28.9
Denmark	1966	11.5	37.5
England	1968	10.0	26.6
Hungary	1968	14.4	14.2
Japan	1969	10.8	27.7
Spain	1968	2.6	31.8
U.S.A.	1973	19.7	35.0

Modified from (Editorial) *Sex education, contraception seen urgent for teenagers as premarital pregnancy illegitimate births increase.* Fam. Plann. Perspect. 7:277, 1975.

TEENAGE PREGNANCY

Extent of Problem

Consider the following statements[5]:

1. In 1973 in the United States there were over 600,000 pregnancies in females under 20.
2. Fifty-three percent of births to single parents in the United States are to adolescents (Table 6-3).
3. Of these, 80 percent are to women of low-income families.
4. The out-of-wedlock birth rate is rising (Figure 6-18). In 1974, 44 percent of all births in the city of St. Louis were to unwed mothers. In 1970 the rate was 32 percent.[6]

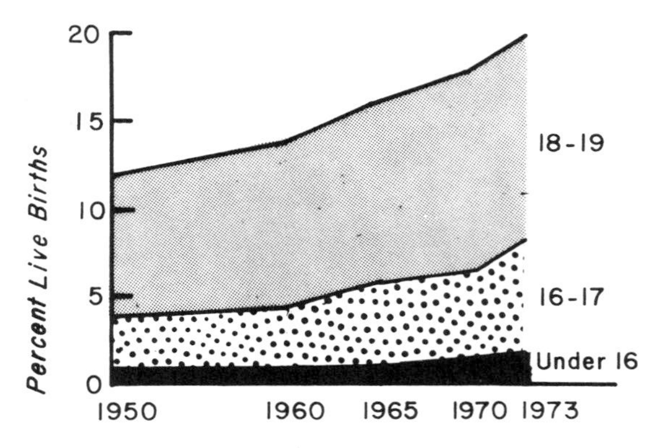

FIGURE 6-19. Percent of live births to women under age 20, U.S., 1950 to 1973. (Adapted from Stickle, G., and Paul, M. A.: Contemp. Obstet. Gynecol. 5:85, 1975.)

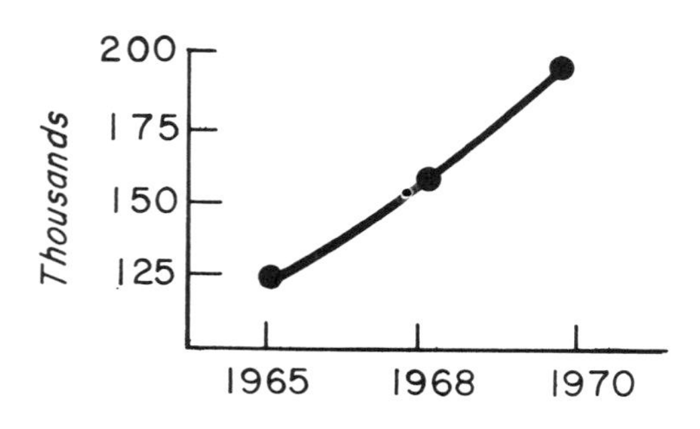

FIGURE 6-18. Number of out-of-wedlock births to teens. (Based on data from Cutright, P.: *Illegitimacy: myths, causes and cures.* Fam. Plann. Perspect. 3:25, 1971.)

5. Since 1950 there has been a 300 percent rise in births involving mothers 15 to 19 (Fig. 6-19), while during that time total births in the United States declined.
6. Thirty-four percent of teenage births were to women still in school.
7. Twenty-five percent of teenage mothers have more than one child before they reach the age of 20 (Fig. 6-20).
8. In England, despite the decrease in the female adolescent population between 1965 and 1971, the number of births per year to this group increased. Moreover, despite decreasing fertility among women aged 20 and over, the fertility of teenagers increased.[7]
9. The decreasing age at menarche (Fig. 6-21), probably related to improved nutritional

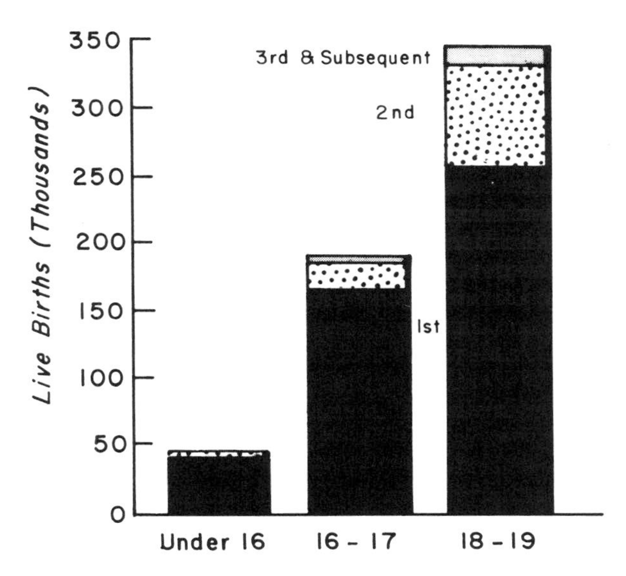

FIGURE 6-20. Subsequent reproduction in adolescent mothers. (Adapted from Stickle, G., and Paul, M. A.: Contemp. Obstet. Gynecol. 5:85, 1975.

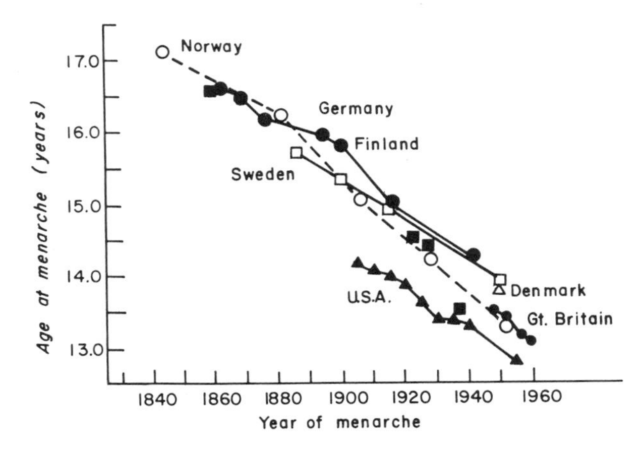

FIGURE 6-21. The declining age of the menarche by decade in several countries. (Adapted from Tanner, J.: *Earlier maturation in man.* Sci. Am. 218:21, 1968.)

status, has increased the proportion of young girls able to conceive.

10. Improved health conditions have increased the chance that a sexually-active girl will conceive and that an out-of-wedlock birth will go to term.

Social Consequences

Sometimes the social consequences of an adolescent pregnancy are not evident immediately. Some pregnancies have long-range consequences. Study the following sampling and note that some problems lead to the next problem.

1. Health threat to mother
2. Leading cause of school dropout
3. Inability to develop marketable skill
4. Low income can be expected
5. Increased welfare dependency
6. Sudden burden of responsibility and social confinement without emotional or financial preparation
7. Adolescent divorce rate twice that of later occurring marriages
8. Shattered lives
9. Disruption of the quality of that individual's life
10. Increased suicide rate in population of females pregnant as adolescents

Medical Consequences

Pregnancy in the younger adolescent means an increased risk of toxemia, anemia, prolonged labor or cephalopelvic disproportion. There is also an increased risk of cholecystitis in teenagers who have been pregnant.

Undernutrition is common among pregnant teenagers; this is secondary either to poverty or teenage faddism. In those women who were pregnant as teenagers there is subsequent higher parity and shorter birth intervals.[8]

Consequences to the Child

In addition to the consequences of teenage pregnancy on the mother, there are also profound consequences which relate to the child.[9]

First, there is a decided increase in infant mortality. Note in Figure 6-22 the increase in number of deaths to infants ranging from the 24-year old mother on the left to the under-15-year old mother on the right.

There is an increase in premature and low birth weight infants (Fig. 6-23). This is as-

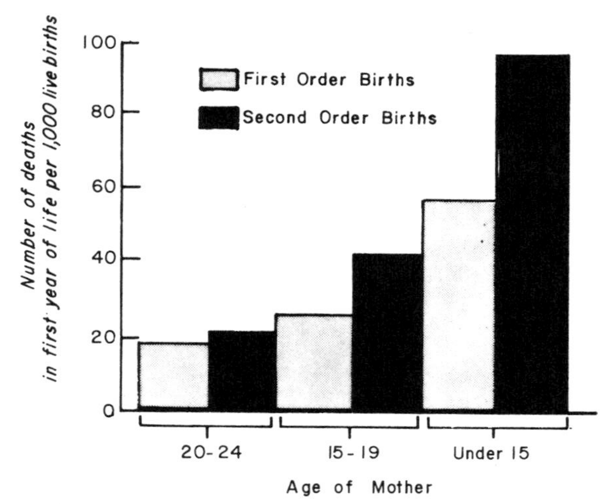

FIGURE 6-22. Mortality per 1000 live births by age of mother and first and second child, U.S., 1969. (Adapted from Whelan, E. M., and Higgins, G. K.: *Teenage Childbearing: Extent and Consequence.* Consortium on Early Childbearing and Childrearing, Child Welfare League of America, Inc., January, 1973. p. 19.)

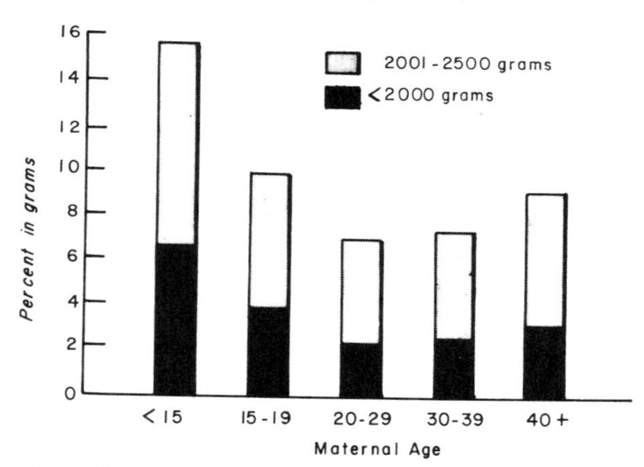

FIGURE 6-23. Percent of liveborn neonates weighing 2500 grams or less by maternal age, U.S., 1973. (Adapted from Stickle, G., and Paul, M. A.: Contemp. Obstet. Gynecol. 5:85, 1975.)

sociated with a seventeenfold increase in the infant death rate, in subsequent lower intelligence quotient scores (Fig. 6-24) and in poorer perceptual functions and motor coordination.

Physical growth within the first year of life is below the national norm, and there is subsequent emotional trauma to intellectual and

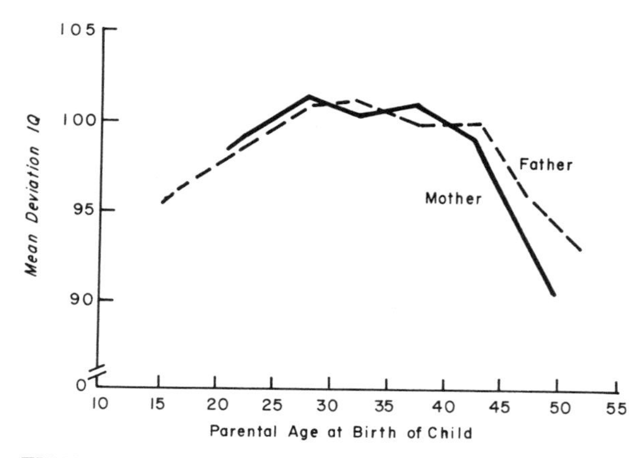

FIGURE 6-24. Mean deviation IQ, children aged 6 to 11, by parental age at birth of child. (Adapted from Stickle, G., and Paul, M. A.: Contemp. Obstet. Gynecol. 5:85, 1975.)

social growth and development associated with fatherless homes and ill-prepared parents.

Specialized Facilities

The multiple needs of the pregnant adolescent must be met in order to provide that adolescent and the infant with optimum care. These needs to be met include medical, psychological, religious, social and academic. Specialized comprehensive obstetrical facilities have been developed to deal with adolescents and to meet their needs. In these facilities there should be in service a variety of health care professionals such as obstetrician, pediatrician, physician's assistant (and/or nurse clinician, a new professional family planning worker), nurse, social worker, teacher and public health worker.

The purpose of these facilities are the following:

1. Provision of a special facility to care for all of the special needs of the adolescent at this critical time.
2. Preparation for childbirth—both medical and psychological.
3. Avoidance of the isolating climate of a large obstetrical clinic with older patients.
4. Provision of comprehensive obstetrical care with a resultant decrease in the medical problems listed previously.

5. Opportunity for completion of education in order to become more productive citizens and better parents.
   a. Completion of high school will increase by a factor of 5 the opportunity for employment in this group.
   b. High school graduates have fewer children than dropouts on a nationwide average.
6. Encouragement and development of a feeling of self-esteem in these young mothers.
7. Development of good habits in preventive medicine for mother and child.
8. Provision of family planning follow-up.

SUMMARY

As a result of society's unwillingness to face the realities of adolescent sexuality, the pregnancy rate among adolescents is increasing. Many adolescents fail to use family planning measures partly because of ignorance of the menstrual cycle and the fertile period and partly because of the type of contraceptive used. Unmarried adolescent females are hesitant to approach a physician for information on fertility control and, therefore, over-the-counter methods are used. The consequences of adolescent pregnancy are considerable, ranging from social to psychological to medical, involving both the mother (or mother and father) and the newborn, and sometimes resulting in suicide. Specialized comprehensive obstetrical facilities have been developed to meet the needs of pregnant adolescents, thus serving both present needs and future opportunities.

QUESTIONS

1. What are the sexual patterns of adolescents?
2. What are the contraceptive habits of adolescents?
3. What are the disadvantages of adolescent pregnancy?
4. How can unwanted births among adolescents be prevented?

REFERENCES

1. Kantner, J. F., and Zelnick, M.: *Sexual experience of young unmarried women in the United States.* Fam. Plann. Perspect. 4:1, 1972.
2. Kantner, J. F., and Zelnick, M.: *Contraception and pregnancy: experience of young unmarried women in the United States.* Fam. Plann. Perspect. 5:24, 1973.
3. Klein, L.: *Early teenage pregnancy, contraception and repeat pregnancy.* Am. J. Obstet. Gynecol. 120:249, 1974.
4. Shaw, F., Zelnick, M., and Kantner, J. F.: *Unprotected intercourse among unwed teenagers.* Fam. Plann. Perspect. 7:39, 1975.
5. *Summary Report: Final natality statistics 1973.* Monthly Vital Statistics Report, National Center for Health Statistics (DHEW), Vol. 23, No. 11, Supplement 1975, p. 7.
6. *The Philadelphia Bulletin.* December 1, 1975, p. 1.
7. Teper, S.: *Recent trends in teenage pregnancy in England and Wales.* J. Biosoc. Sci. 7:141, 1975.
8. Luenhoelter, J. H., Jiminez, J. M., and Baumann, G.: *Pregnancy performance of patients under fifteen years of age.* Obstet. Gynecol. 46:49, 1975.
9. Dott, A. B., and Fort, A. T.: *The effect of maternal demographic factors on infant mortality rates.* Amer. J. Obstet. Gynecol. 123:847, 1975.

BIBLIOGRAPHY

Omran, A. R.: *Health Benefits for mother and child.* World Health. January 1974.

7 Contraceptive Methods

Medical education too often emphasizes therapeutic rather than preventative aspects of disease and the importance of contraception in the field of preventative medicine has been underestimated for years. Consider the following statements:

1. Contraceptive methods usually do not protect against disease.
2. Pregnancy is viewed as a physiologic state and complications of pregnancy have not been viewed realistically relative to the hazards of daily living.
3. The concept of a newborn as being potentially detrimental to family welfare under certain circumstances is not easy to accept.
4. Contraception in the past was generally patient-controlled without requiring medical assistance.

Many physicians are still reluctant to initiate the subject of fertility control with their patients. This is true for a variety of reasons such as the physician's moral or religious attitudes, his/her concept of the patient's moral or religious attitudes, his/her own unfamiliar-

Table 7-1. Circumstances under which physicians discussed family planning with female patients.

	Circumstances of female patient					
Physician's position	*Married with mitral stenosis and 2 children*	*Married with 3 children and 1 bedroom*	*Married with 3 children and no social or health problems*	*18-year old married primipara*	*Unmarried with a baby*	*Bride*
Introduced subject of family planning	90	61	21	38	40	23
Discussed only if asked directly	8	38	74	58	47	74
Did not discuss even when asked directly	1	1	4	3	11	2
Discussed sometimes	1		1	1	2	1
Number of physicians	1383	1369	1385	1384	1375	1383

Data from Peel, J., and Potts, M.: *Textbook of Contraceptive Practice*. Cambridge University Press, Cambridge. 1970, p. 249.
Note: Data resulting from an inquiry among 1989 physicians in England and Wales. These physicians are from the same area as those in the retrospective survey related to the pill and thromboembolic phenomena, given later in this chapter.

ity with the subject, the patient's embarassment on discussing the subject and general criticism of previous medical contraceptive practices and attitudes. Physicians initiate the offer of the use of a first medical contraceptive method in only a third of their patients.[1]

A 1971 survey[2] in three urban areas and one rural community revealed that (1) 75 percent of physicians would initiate contraceptive discussions with a married patient with several children or during a premarital examination; (2) Only 40 percent would initiate contraceptive consultation with a childless sexually-active minor; and (3) 26 percent would refuse to prescribe contraception even if the minor requested it.

A further indication of the circumstances under which physicians discussed family planning with female patients is shown in Table 7-1.

Reluctance on the part of the physician to initiate this discussion runs counter to the patient's strong desire for the physician to take the initial step. This is emphasized in the following studies:

1. In North Carolina, 88 percent of college students interviewed in 1968 and 1969 desired that the physician provide contraception counseling at the premarital exam.[3]

2. In Baltimore, 92 percent of low income black families believed the physician should inquire if a patient is interested in contraception.[4]

3. In Philadelphia, 86 percent of women in their third trimester who were going to a university obstetrical clinic expressed a desire to learn about family planning.[5] Many of these patients were hesitant to discuss the subject with clinical personnel because they believed "the doctor was in too much of a hurry" or "the doctor wouldn't know what to tell them anyway."

Despite this strong desire of patients to be provided with the means to control their fertility and despite the improvements in contraceptive technology over the past decade, the following[6] must be considered:

1. 44 percent of all births to currently married females were unplanned.
2. While only 1 percent of first births were "never wanted," nearly two thirds of all sixth or higher births were so reported.

As indicated previously in Chapters 1, 5 and 6, there is an inverse relationship between the education of a female and the occurrence of unplanned and unwanted pregnancies (Fig. 7-1).

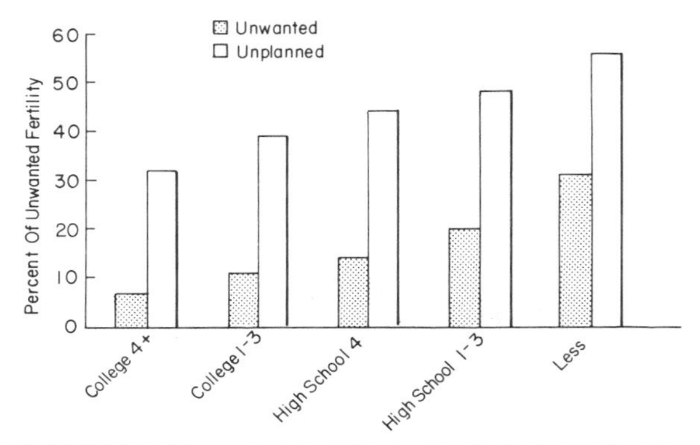

FIGURE 7-1. Unwanted fertility in the U. S. in 1970 with respect to education of the female. (Based on data from *The Report of the Commission on Population Growth and the American Future,* March 1972.)

EFFECTIVENESS OF CONTRACEPTIVE UTILIZATION

It is difficult to evaluate the effect of contraceptive utilization on the fertility rate of a particular group. This is partly because there are numerous demographic forces that influence fertility as mentioned in previous chapters. (The marked decline in nineteenth century European fertility was without the benefit of efficient contraception.) Nonetheless, the following have been noted:

1. Areas with higher rates of contraception acceptance or more program activity tend to have larger decreases in fertility independent of socioeconomic change. For example, when seven counties in rural Georgia with family planning programs were compared with matching counties lacking such programs it was discovered that there was a 30 percent decline in general fertility in the counties with

family planning programs and only 17 percent in the others.[7]

2. The decline in fertility among contraceptive acceptors is greater than among non-acceptors in a matched population. For instance, a comparison of postpartum contraception acceptors in Washington, D.C., matched for age, parity and marital status with non-acceptors revealed a 58 percent lower-than-expected pregnancy rate during the 12 months following delivery in the former, with only a 7 percent lower rate in the latter.[8]

3. Acceptors in family planning programs experience lower fertility after acceptance than before. Excluding the effects of aging, normal postpartum infertility and abortion, there was a 56 percent decline from pre- to post-acceptance fertility noted in an international postpartum program sponsored by The Population Council.[9] Pregnancy rates established in The Population Council annual report further illustrate this point (Fig. 7-2).

4. Many countries without active family planning programs fail to show fertility de-

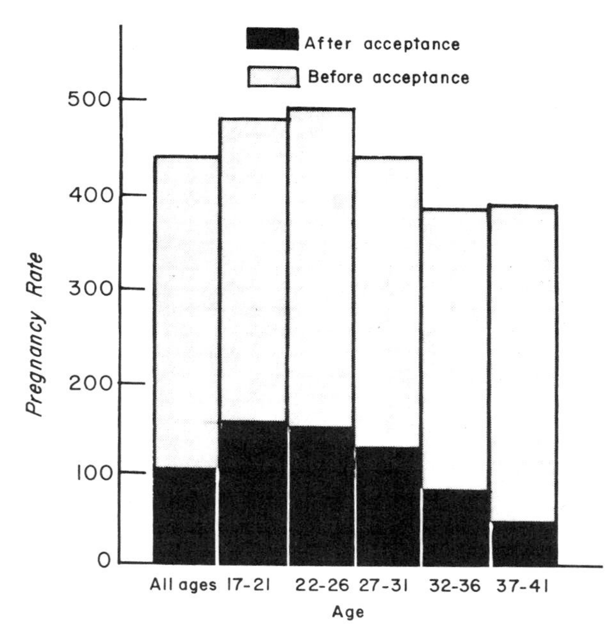

FIGURE 7-2. Pregnancy rates before and after acceptance in the postpartum program, by age. (Adapted from Berenson, M.: Population Council Annual Report, 1971.)

Table 7-2. Estimated contraceptive usage based on sample surveys of fertile couples in selected developed and developing countries, 1967 to 1972.

Country	*Percent using birth control*	*Method of birth control used*					
		Pills	*IUDs*	*Condoms*	*Rhythm or withdrawal*	*Other*	*Total*
Developed							
Australia	92	38	8	9	33	12	100
Japan	52	2		68	41	30	141
Sweden	73	14	1	38	21	11	85
United Kingdom	87	22	5	37	33	14	111
United States	65	34	7	14	9	36	100
Developing							
Columbia	18	55	31	3	N/A	11	100
Iran	51	41	6	18	35		100
Jamaica	42	26	14	41	12	7	100
Korea	45	31	26	20	20	2	100
Panama	46	59	15	4	13	9	100
Philippines	33	24	9	6	61		100
Thailand	40	60	23	5	5	8	101
Venezuela	48	27	27	19	23	4	100

Adapted from data from Dalsimer, I. A., Piotrow, P. T., and Dunn, J. J.: *Barrier methods.* Population Report, Series H, No. 1, December, 1973, p. 34.

clines despite improvements in social and economic development.[10] Examples are Kuwait, Mexico and Egypt.

5. The dramatic rise in pill use during the 1960's has coincided with the largest decrease yet reported for American fertility.[11]

6. A world survey reveals that areas of low fertility are most often those with high contraception utilization.[12] The impact of this is shown in Table 7-2 and Figure 7-3.

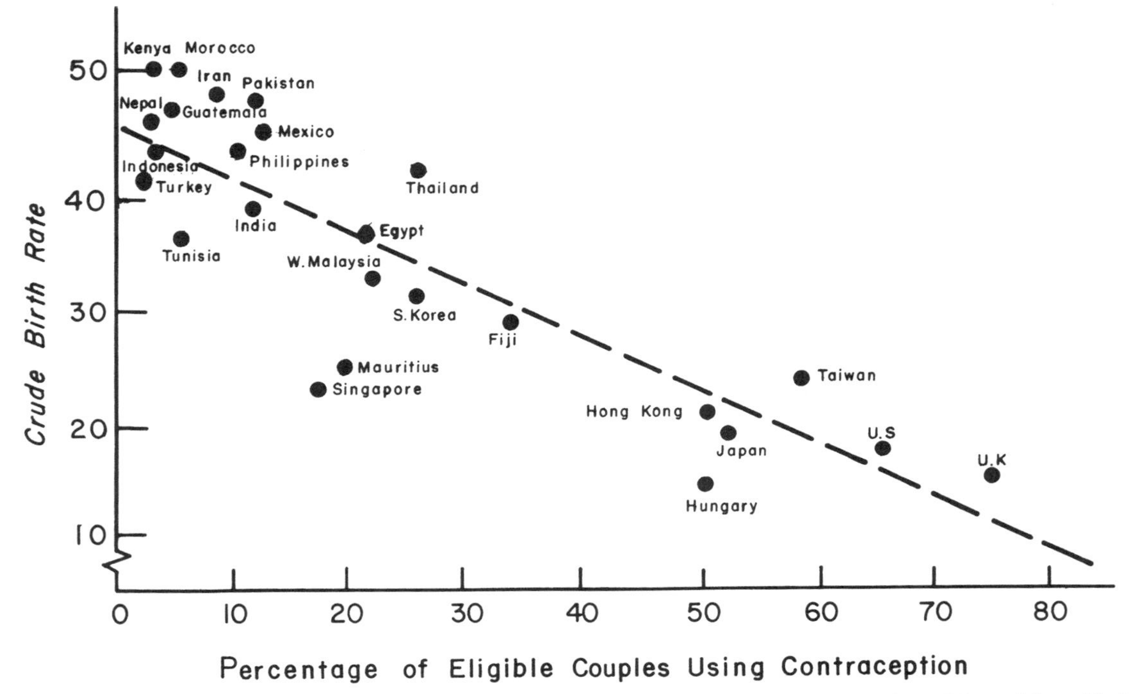

FIGURE 7-3. Comparison of the use of fertility control with the birth rates for selected countries. (Adapted from Nortman, D.: *Population and family planning programs: a factbook*. Reports on Population/Family Planning, No. 2, ed. 5, September 1973.)

Table 7-3. Mean coital frequency in four weeks prior to interview among women currently using contraception by age and type of contraceptive method used, 1970.

Type of method	*Mean frequency by age*			
	All ages	*25*	*25–34*	*35–44*
Total users	8.8	11.1	9.1	6.7
Coitus—independent*	9.3	11.5	9.2	7.2
Coitus—dependent†	8.2	10.4	9.0	6.1
Coitus—inhibiting‡	7.5	8.8	8.5	6.3
Female	9.1	11.3	9.1	6.7
Male	8.4	10.2	9.3	6.8
Couple	7.1	8.6	8.2	5.9

Based on data from Westoff, C. F.: *The modernization of U.S. contraceptive practice*. Fam. Plann. Perspect. 4:9, 1972.

*Coitus—independent: IUD, pill, sterilization
†Coitus—dependent: Condom, foam, diaphragm
‡Coitus—inhibiting: Withdrawal, rhythm

Quality of Life

In addition to the health benefits of decreased fertility, which we referred to in previous chapters, one must consider effects of contraception on the quality of the patient's life. The three statements which follow pertain to the effect of contraceptives on sexual relationships, as one sampling of the quality of life.

1. In the period between 1965 and 1970 there was an increase in coital frequency among married females and their husbands (Fig. 7-4).
2. The highest frequency of coitus is associated with the most effective methods; the proportion of females utilizing these methods has increased (Fig. 7-5, Table 7-3).
3. The highest frequency of coital activity is also noted between couples who wish to delay rather than prevent pregnancy (Table 7-4).

Although the overall proportion of females using contraception varied little (63.9 percent in 1965 to 65 percent in 1970), the patterns of utilization have changed significantly (see Fig. 7-5). Consider the following statements:

1. Voluntary sterilization is now the most popular method of contraception currently used by older couples in which the wife is between 30 and 44, and the second most common contraceptive modality among all married American couples.[13]

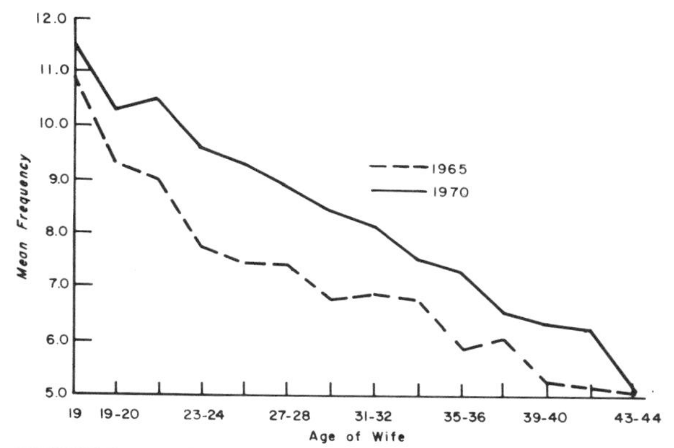

FIGURE 7-4. Mean coital frequency in four weeks prior to interview by age, 1965 and 1970. (Adapted from Westoff, C.F.: *Coital frequency and contraception.* Fam. Plann. Perspect. 6:136, 1974.)

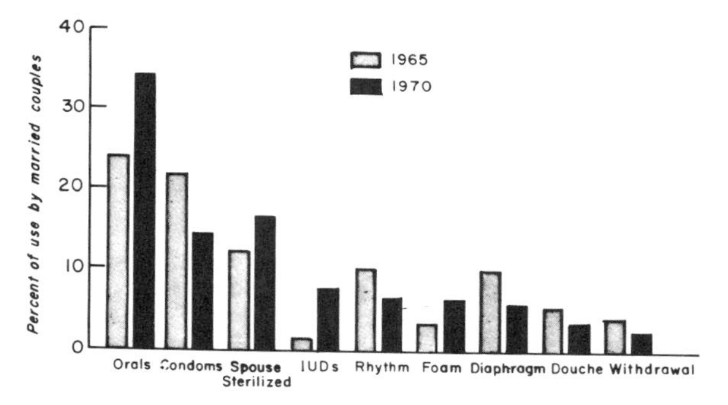

FIGURE 7-5. Change in U. S. contraceptive usage, 1965 to 1972.

Table 7-4. Mean coital frequency in four weeks prior to interview among women currently using contraception, by age, according to whether the intention is to delay a wanted birth or to prevent an unwanted birth, 1970.

Future fertility intention	*Mean frequency, all ages*			
	All ages	*25*	*25–34*	*35–44*
Total users	8.8	11.1	9.0	6.7
Delay	10.6	11.6	9.5	7.7
Prevention	8.0	9.8	8.9	6.6

Based on data from Westoff, C. F.: *The modernization of U.S. contraceptive practice.* Fam. Plann. Perspect. 4:9, 1972.

2. The pill is the most popular method utilized by 34.2 percent of couples in 1970 versus 23.9 percent in 1965 (Fig. 7-6).

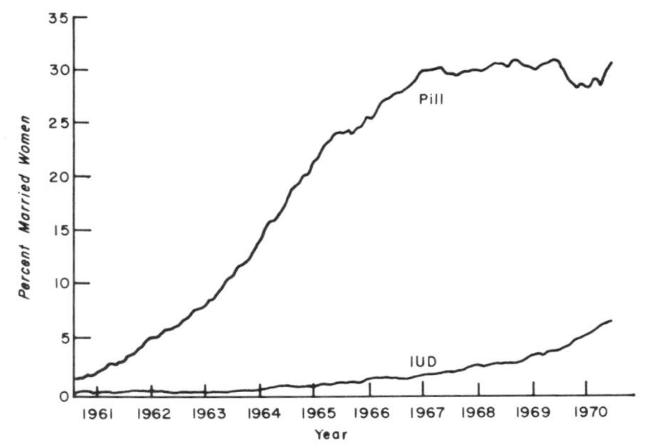

FIGURE 7-6. Pill and IUD usage among married women under 35 from January 1961 through September 1970. (Adapted from Ryder, N. B.: *Time series of pill and IUD use: United States 1961–1970*. Stud. Fam. Plann. 3:234, 1972.)

3. Although there has been an increase in the use of medically administered contraception, the utilization of less efficient contraception remains a problem among sexually-active unmarried young females (see Chapter 6).

4. Certain feminist groups have felt that the hazards associated with less effective contraception coupled with early abortion backup (menstrual regulation) are less than the alleged medical hazards associated with the more effective methods.

Contraceptive Failure

Of the many pertinent considerations in contraception provision, the health care worker must understand the factors which pertain to contraceptive efficacy. In assessing the continuation rates of a particular contraceptive, that is, the duration of time that method was used before it was stopped, the various influences on this rate must be considered. Such influences are age; parity; availability of method; motivation; political, philosophical, economic, cultural and religious factors; sexuality; need; side effects; and health facilities in addition to the specific intrinsic qualities of the contraceptive itself. Continuation rates thus vary greatly for different individuals in different countries and in various clinical situations.[14] Figure 7-7 shows how a variable such as parity can affect the selection and continual use of a particular method.

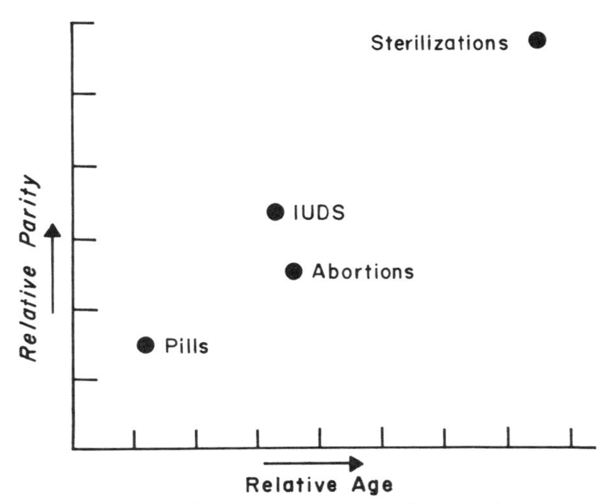

FIGURE 7-7. Median parity versus median age for various contraceptives. (Adapted from Potts, M.: *Introduction: problems and strategy*. In Potts, M. and Wood, C. (eds.): *New Concepts in Contraception*. University Park Press, Baltimore, 1972.)

"Continuity of use can be critical; for example, an annual acceptance rate of 5 percent of nonusers with 30 percent discontinuing each year (e.g. the IUD) would stabilize at about 15 percent current use and four points off the birth rate, whereas the same 5 percent discontinuing (e.g. sterilization, with discontinuation referring to older users aging out of the period of fecundity) would take 10 points off the birth rate in about a decade—and it would take only 4 years to take ten points off the birth rate with twice the acceptors."[15] There is no doubt, however, that an improved dialogue between the contraceptive administrator and the acceptor will contribute to improved compliance.

As mentioned previously, when assessing contraceptive efficiency, the health worker should be aware of the failure rates that can occur with each method. These can be evaluated in terms of the (1) theoretical effectiveness of the method or the failure rate per 100 female years of use by constant user, implying perfect usage without error or omission, and (2) use-effectiveness of the method or the failure rate among actual users of that contraceptive.

Examples of the patient's contribution in altering use-effectiveness are the patient's forgetting to take her pills, not checking for the IUD strings, and not using the diaphragm she was given.

One must also consider the extended use-effectiveness of a particular method. For instance, a patient requests that her IUD be removed. In the interim, prior to institution of a new method, she becomes pregnant. This would be considered a method failure for the IUD.

Contraception effectiveness has been evaluated in terms of the number of pregnancies per one hundred women years (HWY) of use of that method. These failure rates can be derived by application of the Pearl formula: the number of pregnancies per women years = 1300 times the total number of failures divided by the total number of cycles under consideration. As an example[16]:

	Effectiveness	
	Theoretical	Use
Pill	0.1	0.7
IUD	1.9	2.8
Condom	2.6	17.0
Diaphragm	2.5	18.0
Withdrawal	16.0	23.0
Rhythm	14.0	40.0

The Pearl formula unfortunately does not take into account the length of time the method has been practiced. Higher failure rates of short-term users may be overrepresented. Life table analysis can be used instead to measure continued use over a specific interval of time.

Despite a 50 percent improvement in contraception success since the 1950's there is still a 34 percent failure rate in contraceptive utilization among married American females. *Contraceptive failure reflects the characteristics of the patients who use each method as well as the method itself.*

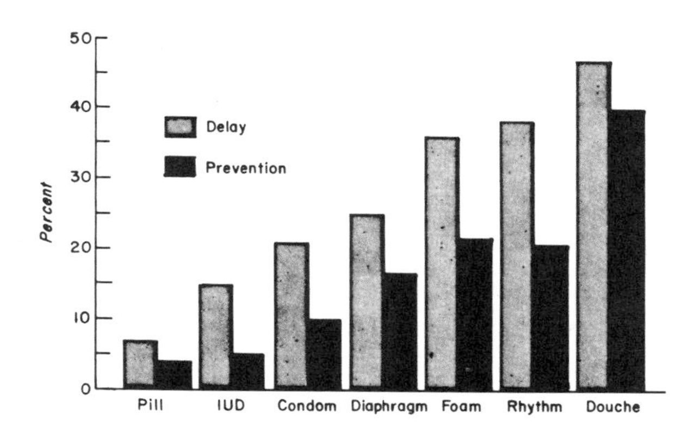

FIGURE 7-8. Percent of contraceptive users who fail to delay a wanted pregnancy or to prevent an unwanted pregnancy in the first year of exposure to risk of unwanted conception. (Adapted from Ryder, N. B.: *Contraceptive failure in the U. S.* Fam. Plann. Perspect. 5:133, 1973.)

Table 7-5. U.S. contraceptive failure as per method per hundred women years (HWY).

Method	*Couples intending to delay next pregnancy*	*Couples intending to prevent further pregnancies*
Pill	7	4
IUD	15	5
Condom	21	10
Diaphragm	25	17
Foam	36	22
Rhythm	38	21
Douche	47	40
Overall rates, all methods	26	14
Nothing	90+ (estimate)	

Adapted from Ryder, N. B.: *Contraceptive failure in the United States.* Fam. Plann. Perspect. 5:133, 1973.

A study[17] of contraceptive failure among ever married females in 1970 revealed:

1. The rate of failure is higher if the intention is to delay rather than to prevent pregnancy (Fig. 7-8 and Table 7-5). Note that delay refers to child spacing and prevention refers to a discontinuation of further reproductive function.
2. A lower proportion of older females experience contraceptive failure than younger patients.
3. There is no systematic variation in contraceptive failure by pregnancy order or by the wife's education.
4. Blacks appear much less successful than whites both in delaying the next wanted pregnancy and in preventing the next unwanted pregnancy.
5. White non-Catholics are more successful than white Catholics in attempting to delay the next pregnancy, but both groups are equal in their attempt to prevent the next unwanted pregnancy.
6. The percentage of females attempting both to delay and to prevent pregnancy is higher in females of later marriage cohort (1966–1970) reflecting a change in attitudes mentioned in previous chapters.
7. The most important factor in the decline of failures is attributed to the adoption of the pill.
8. As discussed in Chapter 6, choice of a medically administered contraceptive (pill, IUD, diaphragm) increased by a factor of four the utilization of the method (Table 7-6).

Table 7-6. Ratio of users to non-users according to type of contraceptive used at last intercourse.

Method used	*Ratio*
Pill, IUD, diaphragm	3.7
Condoms, foam	1.1
Withdrawal	0.6
Douche, rhythm	0.4

Data adapted from Ryder, N. B.: *Contraceptive failure in the United States.* Fam. Plann. Perspect. 5:133, 1973.

In trying to understand and prevent these contraceptive failures one must be aware that

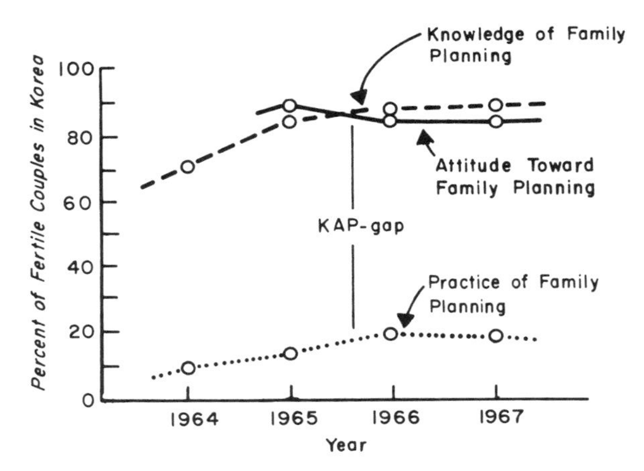

FIGURE 7-9. The KAP-gap in Korea. (Adapted from Rogers, E.: *Communication Strategies for Family Planning.* Free Press, Macmillan Co., New York, 1973, p. 183.)

there is frequently a large gap in the patient's *knowledge of, attitude toward,* and *practice of* family planning (Figure 7-9). This is called the KAP-gap. The importance of education, communication and persuasion to enhance patient motivation to reduce this gap cannot be overemphasized.

In explaining the reasons for this KAP-gap one should examine not only the patients'

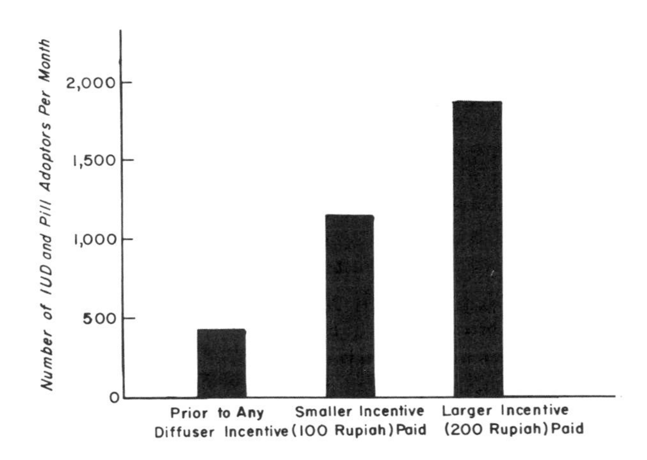

FIGURE 7-10. Initiation of a smaller and a larger diffuser incentive in Indonesia led to an increase in rates of adoption of IUD and pills. (Adapted from Rogers, E.: *Communication Strategies for Family Planning.* Free Press, Macmillan, New York, 1973, p. 288.)

motivations but also (and probably more importantly) the availability of family planning facilities. This availability refers not only to the presence of facilities but also to other important factors such as the hours of operation (e.g. night clinics for working women) or facilities for child care during the patient visit. It has also been demonstrated in many instances that the establishment of incentives either for the providers or the acceptors of family planning can increase the rates of acceptance (Fig. 7-10).

METHODS

With some background on contraception per se, the rest of the chapter will be a discussion of the various methods, their advantages and disadvantages, effectiveness, risk of failure and medical risk. Clinical situations will be presented so the health care worker can assess the problems that he or she might face in dealing with a patient.

Regardless of the method prescribed, the health care worker must *always* keep the following in mind when prescribing a method of fertility control:

1. Assess the patient and her situation
   a. Basic motivation: spacing vs. limitation
   b. Size of family
   c. Sexual pattern and attitudes
   d. Socioeconomic considerations such as:
      (1) degree of privacy
      (2) availability of sanitary facilities
      (3) lack of education
2. Consider the effectiveness in the light of the couple as a "biologic unit."
3. Guide the patient to make the choice according to special and specific needs.
4. Use medical judgment and skill. (Specifically, evaluate general health status of patient.)

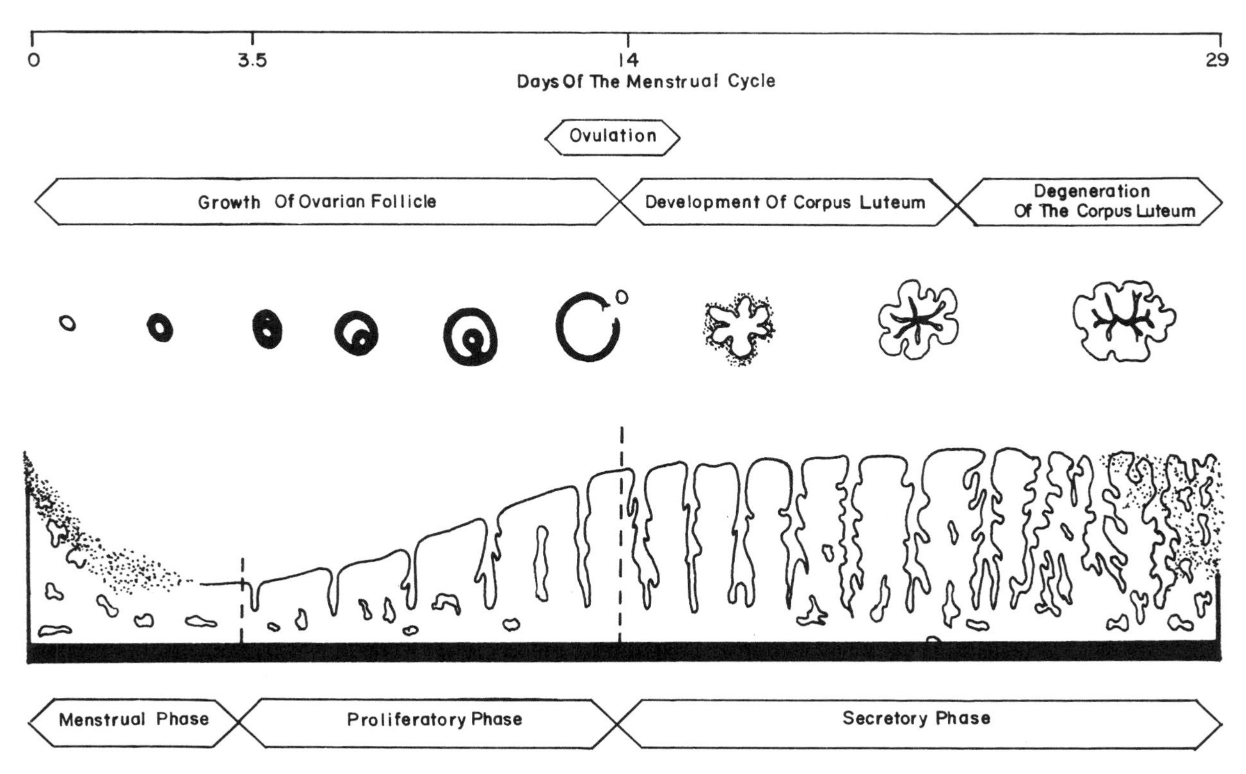

FIGURE 7-11. Ovarian and endometrial activity during a classic menstrual cycle.

Rhythm Method

When a given population is desirous of achieving a pregnancy, there are evidences of the following lengths of delay in achieving conception (in months):

1 mo.	*3 mo.*	*6 mo.*	*12 mo.*
28–40%	58–58%	78–82%	87–93%

The theoretical probability of pregnancy resulting from a single unprotected coitus is 2 to 4 percent.[18] The rhythm method, then, attempts to minimize this risk by eliminating the random exposure. Basically the rhythm method avoids coitus during the fertile phase of the cycle.

Although ovulation generally occurs 14 days prior to menstruation, there is considerable variation in the proliferative or preovulatory phase of even the most regular patient (Fig. 7-11). In reviewing this figure, remember the time interval during the growth of the ovarian follicle can be quite variable. This means the time from the previous menstrual period to ovulation cannot be safely predicted.

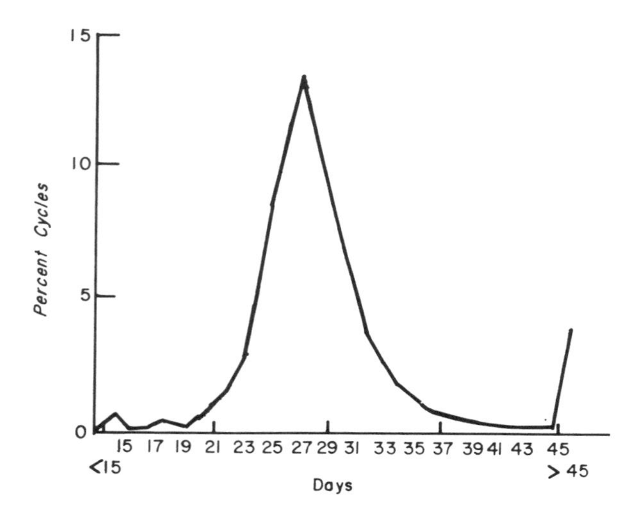

FIGURE 7-12. Distribution of cycle length based on a study of 30.655 cycles in 2316 regularly menstruating women of all ages. (Based on data from Chiazze, L., Jr., et al.: *The length and variability of the human menstrual cycle.* J.A.M.A. 203:377, 1963.)

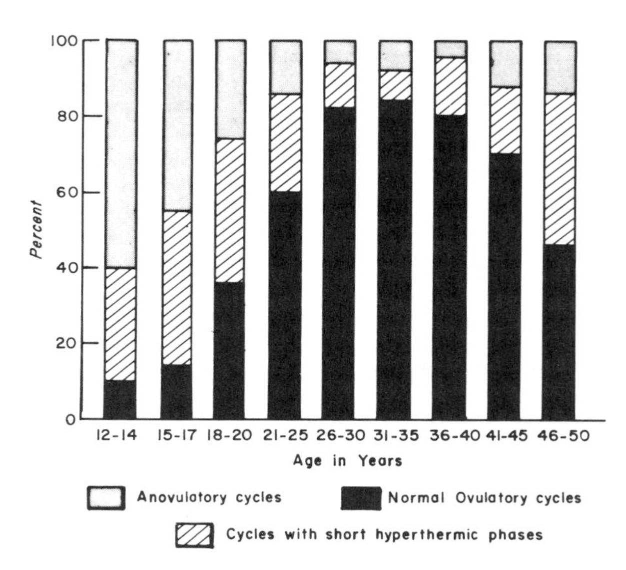

FIGURE 7-13. Distribution of the ovulatory cycle, cycles with short hyperthermic (luteal) phases, and anovulatory cycles in women from the age of 12 to 50 years. Note the large percentage of nonovulatory cycles in the pubertal and immediately postpubertal females. (Adapted from *Biology of Fertility Control by Periodic Abstinence.* World Health Organization Technical Series, No. 300, Geneva, Switzerland, 1967.)

Moreover, there is a significant percentage of anovulatory cycles at all ages (Fig. 7-12). Also, it should be emphasized that variability increases at extremes of the reproductive years (Fig. 7-13).

The ancient Hebrews, recognizing this fact of reproductive biology, used the rhythm method in reverse to promote fertility. They prohibited intercourse from the onset of the menses to seven days after its cessation (Leviticus), thereby favoring more fertile times of the cycle.

Problem: How do you advise your patient as to the "safe" interval for coital activity? Keep in mind that viable sperm have been noted in the reproductive tract 5 to 6 days after an isolated coitus and that regularly cycling women may ovulate as early as day 8 to 9 and in rare circumstances even earlier.

Answer: There are three ways for the patient to improve her timing:

1. The most commonly utilized is the recording of the basal body temperature during the

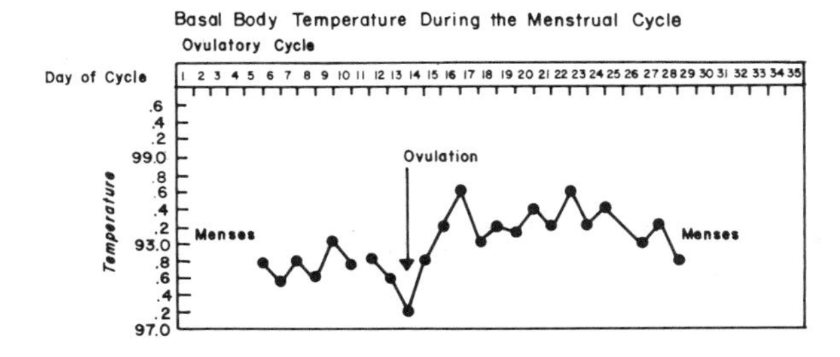

FIGURE 7-14. A typical biphasic body temperature during an ovulatory menstrual cycle.

menstrual cycle (Fig. 7-14). The patient is instructed to record her temperature every morning immediately upon arising and before performing any of her daily routine. Special thermometers are available to facilitate the reading. There is a rise of the patient's temperature following ovulation as a result of progesterone and some of its metabolites being thermogenic, acting in the hypothalamic thermoregulatory centers possibly with norepinephrine or other catecholamines serving as a mediator. Although there is always uncertainty as to the length of the preovulatory safe period, the patient can be assured better relative protection after the third day following the rise in the basal body temperature. The patient should always be wary of the occurrence of colds, infections, etc. which may influence the temperature.

2. Examination of the cervical mucus by the patient during the menstrual cycle has been used to facilitate the decision as to a safe period. The biological characteristics of cervical mucus are depicted in Table 7-7. Errors can occur in interpretation because it is subjective and deviations in the appearance of the cervical mucus can be found with mild cervical infections.

3. Luteinizing hormone releasing factor (LRH) or an analogue have been proposed to regulate ovulation, thus reducing the randomness. This is highly speculative and has not as yet been substantiated by recent studies with currently available analogues.

Summary

Effectiveness: reliable if practiced correctly.

Advantages:

1. No prescription or complicated equipment needed.

2. Sanctioned by the Roman Catholic Church. *Note:* The rhythm method is utilized by only 14 percent of Catholics, while the utilization of all contraception increased from 30 percent in 1955 to 68 percent in 1970 among married white Catholics (especially among the younger cohort aged 20 to 24—78 percent).[19]

Disadvantages:

1. Requires motivation and intelligence. (Refer to section on poor understanding of the menstrual cycle in Chapter on Adolescent Sexuality.) *Note:* Never assume your patient can read a thermometer; always inquire and check.

2. Often requires prolonged periods of continence.

Risk of failure: low if done correctly but increases with decrease in motivation and variability of the menstrual cycle.

Alleged medical risk: none.

Table 7-7. Changes in cervical mucus during the menstrual cycle.

	Follicular phase		*Midcycle*	*Postovulatory phase*	
Elasticity (Spinnbarkeit) (cm)	1–2	3–4	15–20	10–15	1–3
Volume	Moderate	Continuously increasing	Maximal (mucorrhea)	Decreased	Minimal
Viscosity	High	Moderate	Low	Low	High
Ferning	Absent	Present	Well developed	Minimal or absent	Absent
Simms-Huhner test (quantity and motility of spermatozoa)	Few spermatoza motile	Motility increasing	Large number of highly motile spermatozoa	Motility decreasing	Few spermatozoa

Coitus Interruptus

Coitus interruptus involves the withdrawal of the penis prior to ejaculation. It is possibly the oldest contraceptive method used. It is mentioned in the Bible:

> And Onan knew that the seed should not be his; and it came to pass, when he went in unto his brother's wife, that he spilled it on the ground, lest that he should give seed to his brother. (Genesis 38:9)

Historically coitus interruptus is responsible for the decline in birth rates in eighteenth century Western Europe. Probably the most important thing for those who work with adolescents to remember is withdrawal is frequently the first method used by most sexually-active adolescents.

Problems

The major problems involve
1. reliance on the male's self-control and timing
2. termination of the sexual act with the male orgasm, which is often unsatisfactory for the female
3. danger of premature ejaculation and the pre-ejaculatory release of sperm in urethral secretions.

Problem: A young patient calls your office and wants to know if she can get pregnant if her boyfriend ejaculates on her pubic hair after withdrawal.

Answer: Yes: Deposition of semen on a woman's external sexual organs have resulted in pregnancy, probably related to extension of copious midcycle cervical mucus.

Summary

Effectiveness: fair if practiced with a great deal of self-control.
Advantages: can be used in unexpected coitus or when nothing else is available.
Disadvantages: decreased sexual gratification for both partners.
Risk of failure: moderate.
Alleged medical risk: none.

Postcoital Douche

Douching after coitus is not a contraceptive technique but may be used by some women for hygienic purposes. Douching does not protect against venereal disease and if done too frequently may predispose to vaginal irritation.

Summary

Effectiveness: virtually ineffective. Spermatazoa have been noted in the fallopian tubes 5 minutes after exposure to the cervical mucus.
Advantages: used *after* intercourse, rather than interrupting foreplay.
Disadvantages: requires sanitary facilities and equipment; shortens pleasure of the coital experience because of necessity of immediate use.
Risk of failure: very high.
Alleged medical risk: none.

Prolongation of Lactation

The relative period of infertility following parturition is variable and can be very short (Table 7-8). The percentage of conceptions that

Table 7-8. Conception one or two months after delivery of previous pregnancy.

Race	*Conception in 30 days*	*30–60 days*	*Total conception within 60 days*		*Total studied*
White	46	164	204	5.5%	3,677
Black	49	143	192	5.5%	3,469
Other	3	25	28	7.8%	361
Total			424	5.6%	7,507

Adapted from Sever, J. L.: *Rubella immunization risk postpartum.* J.A.M.A., 217:697, 1971.

occur following the first postpartum period and with no contraception is 53 percent within 3 months of the first menses and 82.5 percent within 12 months.[20]

Breastfeeding can be utilized in prolonging the length of the anovulatory period following parturition. This functions through an inhibition of the release of gonadotropin (Fig. 7-15).

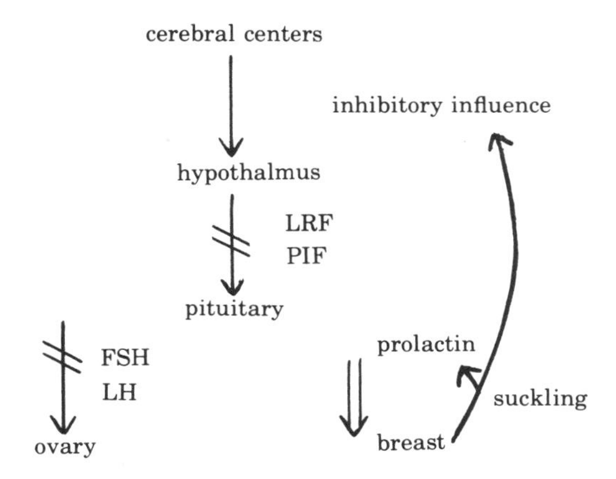

FIGURE 7-15. Effect of lactation on normal function.

This type of protection, however, is relative. The average length of interval from delivery to the first ovulation is 49 days in non-nursing mothers versus 122 days in nursing mothers (Fig. 7-16). Despite the length of this interval, the following was observed in nursing mothers[21,22]: (1) no ovulations were noted in the first 30 days; (2) the risk of ovulation by week 9 was only 1/1250; (3) 78 percent of the first menstruations were ovulatory; (4) effectiveness decreases with supplemental rather than full-time feedings.

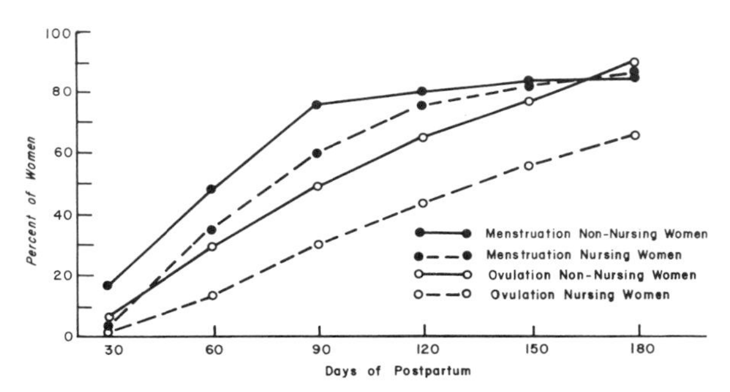

FIGURE 7-16. Return of menstruation and ovulation postpartum. (Adapted from Vorherr, H.: *Contraception after abortion and postpartum*. Am. J. Obstet. Gynecol. 117:1002, 1973.)

Prolongation of lactation is commonly utilized in underdeveloped countries where pregnancy intervals of females who nursed from 1 to 2 years and did not practice contraception were from 5 to 10 months longer than those who did not nurse.

Summary

Effectiveness: brief.

Advantages: requires no medication or planning.

Disadvantages: benefits are short term only; uncertainty of time of return of ovulatory function.

Risk of failure: high.

Alleged medical risk: none.

Antispermicidal Preparations

These preparations include foams, creams, suppositories and jellies that consist of an inert base with a spermicidal chemical which is spread by coital movement. They are used by 6 percent of all United States female contraceptors.

Despite 75 antispermicidal products on the market, there are only 29 active ingredients, most of which are used in combination. There are two basic types: (1) those which do not have surface-active properties, such as organometallic compounds, and especially phenyl mercuric salts; and (2) those which do have surface-active properties. There are three types of the latter—anionic, cationic, and most commonly non-ionic. There are approximately ten bases used. A list of active ingredients is given in Table 7-9.

Failures occur when (1) the preparation is inserted more than 30 minutes prior to coitus; (2) the patient douches within eight hours following coitus; (3) not enough is used; (4) effectiveness of some preparations, e.g. suppositories, is not active until after 15 minutes (Fig. 7-17).

Jellies, Creams & Pastes	Antemin	Clinocap	Conceptrol	Contraceptalene	Delfen Cream	Dura Cream	Dura Gel	Efpa Gel	Genexol	Immolin	Koramex A
Acids											
Boric Acid											●
Ricinoleic Acid											
Tartaric Acid		●									
Bacteriocides											
Benzalkonium Chloride											
Zinc-Phenosulfate											
Phenyl Mercuric Acetate (Nitrate Borate)											●
Chloramine											
Surfactants											
p-Diisobutylphenoxypolyethoxy Ethanol							●		●		
p-Tri-isoprophylphenoxypolyethroxy Ethanol							●	●			
Nonylphenoxy Polyethoxy Ethanol	●		●		●	●	●			●	
Hexylresorcinol		●		●							
Sodium Dioctyl Sulfosuccinate				●		●	●				
Sodium Lauryl Sulfate											
Sodium Dichlorosulfamido Benzoate											
p-Chlormetacresol		●									
Dodecaethylene Glycol											
Polyoxyethylenenonyl Phenol											●
Methoxypolyoxyethylene Glycol Laurate										●	
Polyethyleneglycol Monoisocetyl Phenol Ether											
Polysaccharide-Polysulfuric Acid Ester											
Oxyquinolyne Sulfate											

Jellies, Creams & Pastes (cont.)	Lorophyn	Orthocreme	Orthogynol	Pantentex	Preceptin Gel	Prentif Compound	Ramsey's Contraceptive Jelly	Rendell Cream	Staycept Cream	Staycept Jelly	Volpar Paste
Acids											
Boric Acid		●	●								
Ricinoleic Acid		●	●		●						
Tartaric Acid											
Bacteriocides											
Benzalkonium Chloride	●										
Zinc-Phenosulfate											
Phenyl Mercuric Acetate (Nitrate Borate)	●										●
Chloramine											
Surfactants											
p-Diisobutylphenoxypolyethoxy Ethanol			●		●						
p-Tri-isoprophylphenoxypolyethroxy Ethanol								●	●	●	
Nonylphenoxy Polyethoxy Ethanol				●				●	●	●	
Hexylresorcinol						●					
Sodium Dioctyl Sulfosuccinate						●					
Sodium Lauryl Sulfate		●									
Sodium Dichlorosulfamido Benzoate											
p-Chlormetacresol											
Dodecaethylene Glycol							●				
Polyoxyethylenenonyl Phenol											
Methoxypolyoxyethylene Glycol Laurate											
Polyethyleneglycol Monoisocetyl Phenol Ether											
Polysaccharide-Polysulfuric Acid Ester											
Oxyquinolyne Sulfate											

Suppositories	Clinocap Soluable Suppositories	Lorophyn Suppositories	Orthoforms	Prensols	Prentif Gels	Rendell's Suppositories	Rendell's Wife's Friends Gels	Syn-A-Gen	Staycept Suppositories
Acids									
Boric Acid							●		
Ricinoleic Acid									
Tartaric Acid	●								
Bacteriocides									
Benzalkonium Chloride		●	●						
Zinc-Phenosulfate									
Phenyl Mercuric Acetate (Nitrate Borate)									
Chloramine									
Surfactants									
p-Diisobutylphenoxypolyethoxy Ethanol									
p-Tri-isoprophylphenoxypolyethroxy Ethanol			●						●
Nonylphenoxy Polyethoxy Ethanol			●			●			
Hexylresorcinol				●	●				
Sodium Dioctyl Sulfosuccinate	●			●	●		●		
Sodium Lauryl Sulfate									
Sodium Dichlorosulfamido Benzoate									
p-Chlormetacresol									
Dodecaethylene Glycol									
Polyoxyethylenenonyl Phenol									
Methoxypolyoxyethylene Glycol Laurate									
Polyethyleneglycol Monoisocetyl Phenol Ether		●							
Polysaccharide-Polysulfuric Acid Ester								●	
Oxyquinolyne Sulfate									

	Foaming Tablets: Antibion	Bymeston	Gynomyn	Prentabs	Rendell	Semori	Septon	Areosol Foams: Pantentex	Delfen	Emko	Vaginal Films: C-Film
Acids											
Boric Acid											
Ricinoleic Acid											
Tartaric Acid			●				●				
Bacteriocides											
Benzalkonium Chloride	●										
Zinc-Phenosulfate					●						
Phenyl Mercuric Acetate (Nitrate Borate)											
Chloramine		●	●								
Surfactants											
p-Diisobutylphenoxypolyethoxy Ethanol											
p-Tri-isoprophylphenoxypolyethroxy Ethanol								●			
Nonylphenoxy Polyethoxy Ethanol	●								●	●	●
Hexylresorcinol											
Sodium Dioctyl Sulfosuccinate											
Sodium Lauryl Sulfate											
Sodium Dichlorosulfamido Benzoate				●							
p-Chlormetacresol											
Dodecaethylene Glycol											
Polyoxyethylenenonyl Phenol											
Methoxypolyoxyethylene Glycol Laurate											
Polyethyleneglycol Monoisocetyl Phenol Ether											
Polysaccharide-Polysulfuric Acid Ester											
Oxyquinolyne Sulfate						●					

Table 7-9. Selected vaginal spermicidal formulas.

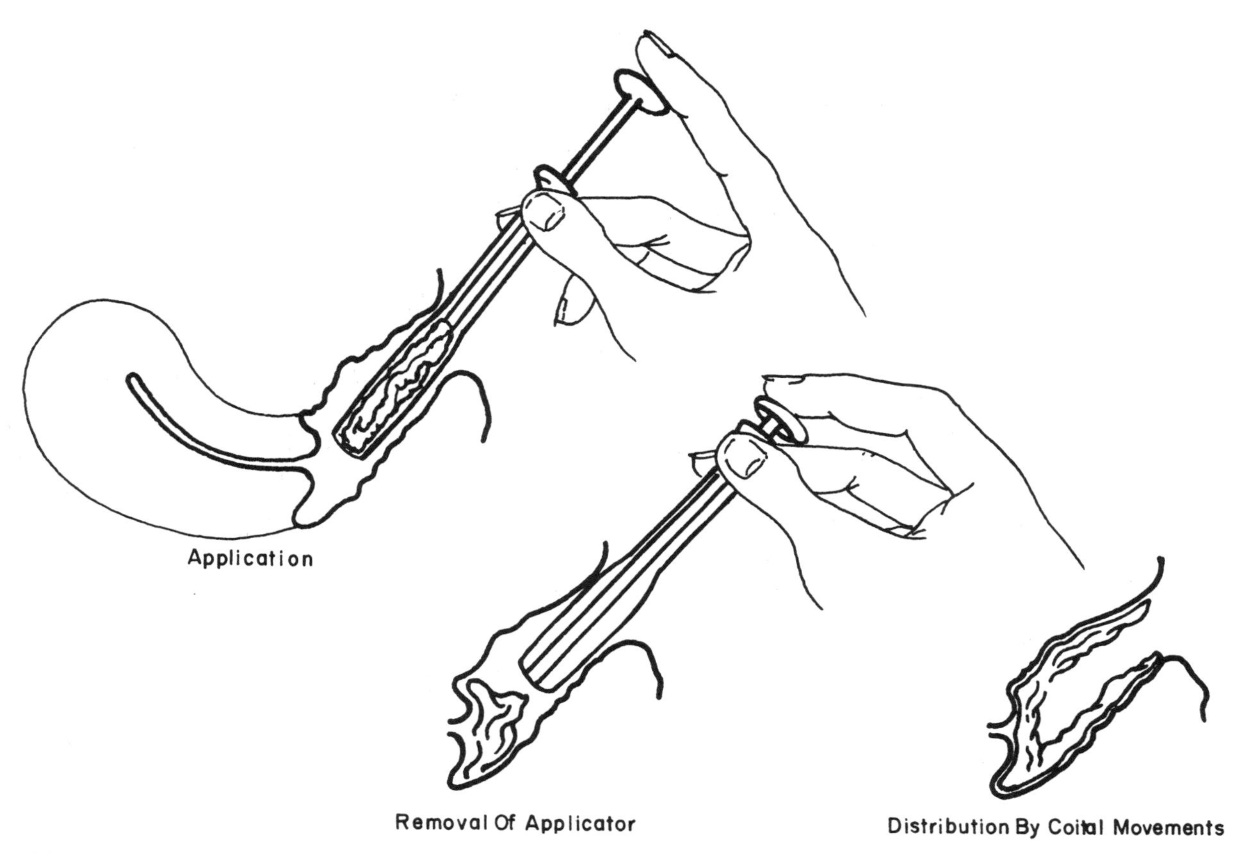

FIGURE 7-17. Application and distribution of spermicidal preparations when used as the sole method of fertility control.

Problem: The patient inquires whether she should apply the preparation more than once a night.

Answer: The preparation should be reapplied with each subsequent coitus.

Summary

Effectiveness: moderately effective.

Advantages: no prescription; may have some protective value against venereal disease; can be used during lactation; can be used as an adjunct to condoms; it is a lubricant.

Disadvantages: strong motivation needed; messy; should not be used prior to orogenital sex.

Risk of failure: relatively low. Five times higher than with oral contraceptives; this relates to inconsistent use rather than method failure.

Alleged medical risks: occasional allergy.

Condom

The condom or "rubber" is a latex sheath, which is sometimes lubricated and comes in various shapes and colors. Occasionally they are made from sheep cecum. It is the second most commonly used contraceptive in the world with over 25 million users worldwide; it is the chief contraceptive in Japan, Sweden and Britain. It is used by 15 percent of contraceptors in the United States, being the third most popular method of contraception. Marked improvements in condom technology, especially in Japan, have diminished the complaint of decrease of sensation. However, an estimated 18 percent of contraceptors in the United States who depend upon the condom fail to delay a wanted pregnancy or to prevent an unwanted pregnancy.[23]

Helpful hints in using the condom are (1) the shelf life of the condom is only about two years; (2) the effectiveness may be increased with concomitant use of foam; (3) the condom can be applied by the sexual partner to add to the act of lovemaking; (4) a space of about 1 cm. should be left at the end as a receptacle; (5) the patient should be instructed to remove the condom immediately following intercourse and to hold onto the base while withdrawing the penis; and (6) lubrication of the condom increases sensations and diminishes irritation and breakage.

Summary

Effectiveness: moderately effective.

Advantages: no prescription; male's responsibility; protection from venereal disease; inexpensive; easy to use; no medical complications; readily available.

Disadvantages: strong motivation needed; occasional breakage; some loss of sensation; precoital interference.

Risk of failure: low.

Alleged medical risk: none.

Diaphragm

The diaphragm is utilized by 4 percent of contracepting couples in the United States or has about 2 to 3 million users. The diaphragm consists of a rubber dome supported by a

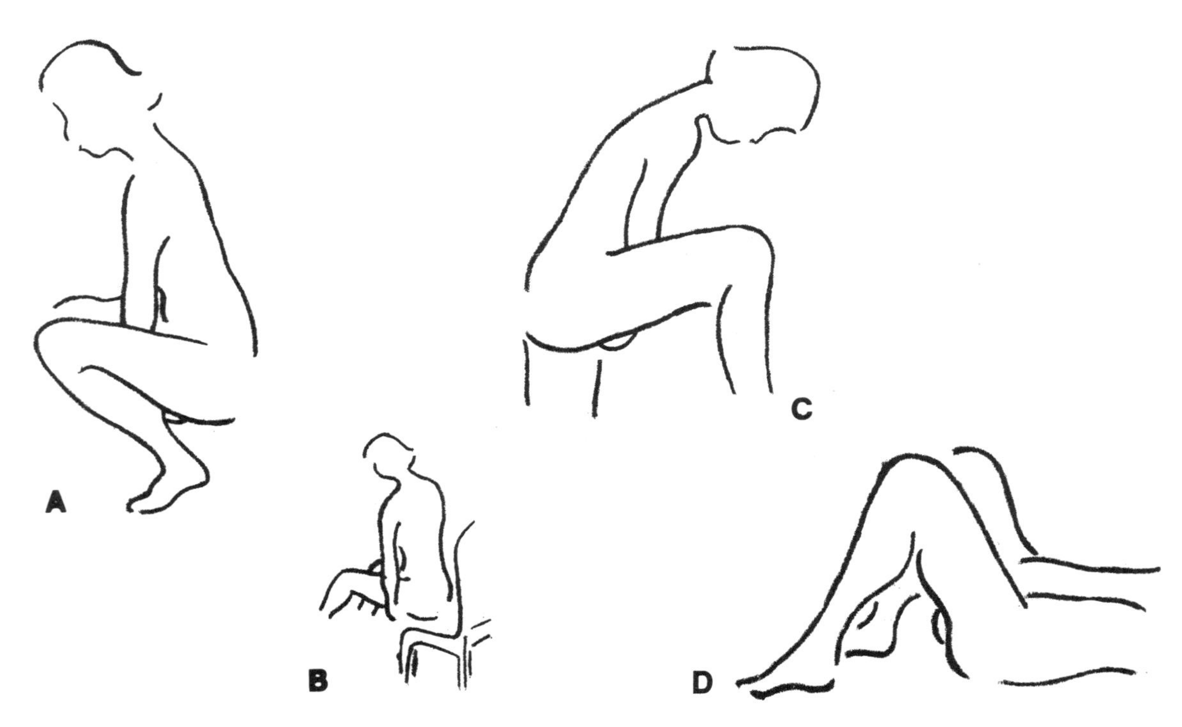

FIGURE 7-18. Positions for inserting the diaphragm. *A, Squatting:* the most frequently used position. *B, Chair method:* position in which the patient sits far forward on the edge of a chair. *C, Standing:* position in which the patient has one foot on a chair or stool. *D, Reclining:* insertion in a semi-reclining position in bed. (From a manual for insertion of the Koromex® diaphragm, prepared by the Holland-Rantos Company, Inc., Piscataway, NJ, with permission.)

rubber-encased metal spring, which fits over the cervix. The size is determined by the distance between the back of the cervix and the pubic bone and is not related to the length of the vagina. There is only a 1 inch difference in size between the largest and smallest diaphragm. Oval shaped diaphragms can be used in special circumstances, e.g. cystocele.

The diaphragm should *always* be used in conjunction with a spermicidal preparation; this greatly enhances its efficacy. The patient should be informed to

1. apply the preparation to both sides of the diaphragm and around the rim prior to insertion. (This will also facilitate insertion by acting as a lubricant.)
2. apply fresh jelly or cream prior to each subsequent coitus without removing the diaphragm.
3. leave the diaphragm in place for six hours following intercourse.

Comfortable positions for inserting the diaphragm are illustrated in Figure 7-18.

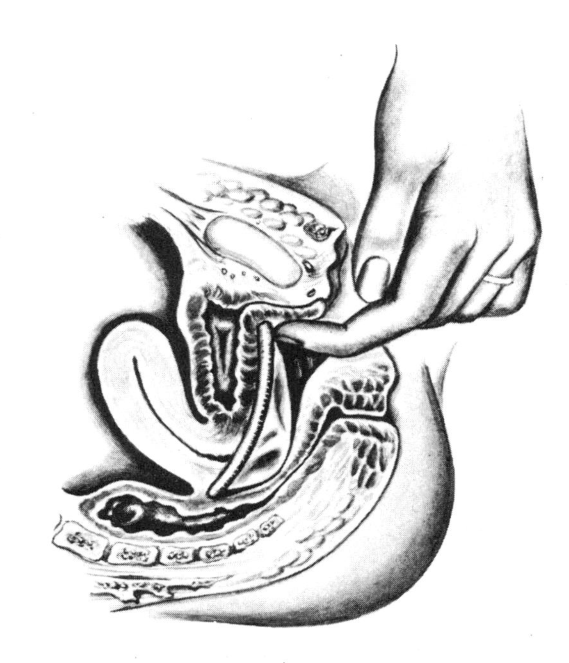

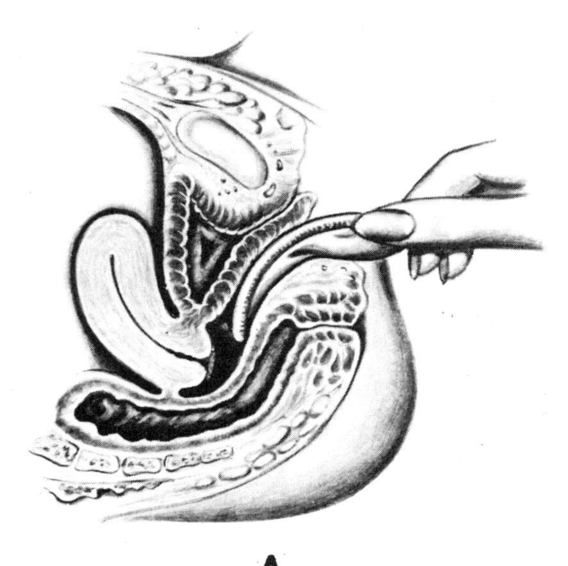

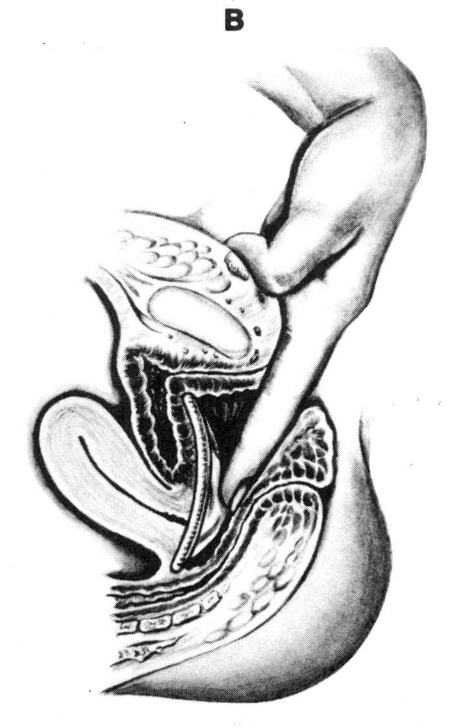

FIGURE 7-19. Steps in the manual insertion of the diaphragm: *A*, Compress the diaphragm, insert in vagina, and push downward as far as possible. *B*, Tip front end upward to fit behind the pubic bone. *C*, Check with finger to make certain it fits over the vagina. (From a manual for insertion of the Koromex® diaphragm, prepared by the Holland-Rantos Company, Inc., Piscataway, NJ, with permission.)

Steps in inserting the diaphragm manually are shown in Figure 7-19. Anteriorly the diaphragm rests against soft tissues posterior to the symphysis pubis; posteriorly the diaphragm lies within the posterior vaginal fornix behind and covering the cervical os; circumferentially the diaphragm is in contact with the vaginal walls. Some women prefer an introducer to facilitate insertion (Fig. 7-20).

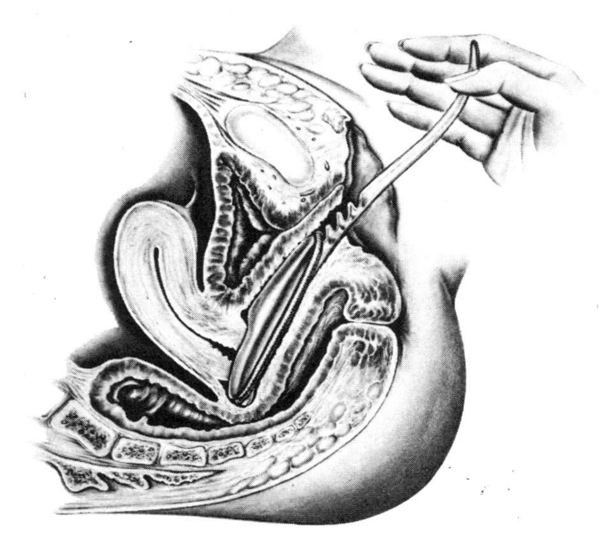

FIGURE 7-20. Using an introducer to insert the diaphragm into the vagina. (From a manual for insertion of the Koromex® diaphragm, prepared by the Holland-Rantos Company, Inc., Piscataway, NJ, with permission.)

If necessary the diaphragm can be worn for prolonged times without discomfort to the wearer. It is easily removed by hooking the index finger into the anterior rim of the diaphragm (Fig. 7-21), pulling down and then out. It can also be removed from the lateral aspect of the vagina if the patient cannot reach under the pubic bone. The diaphragm should be washed with warm water and mild soap after usage and allowed to air dry.

Problem: Can every woman wear the diaphragm?

Answer: No. Variations in pelvic anatomy such as a cystocele or uterine prolapse can prevent proper fitting.

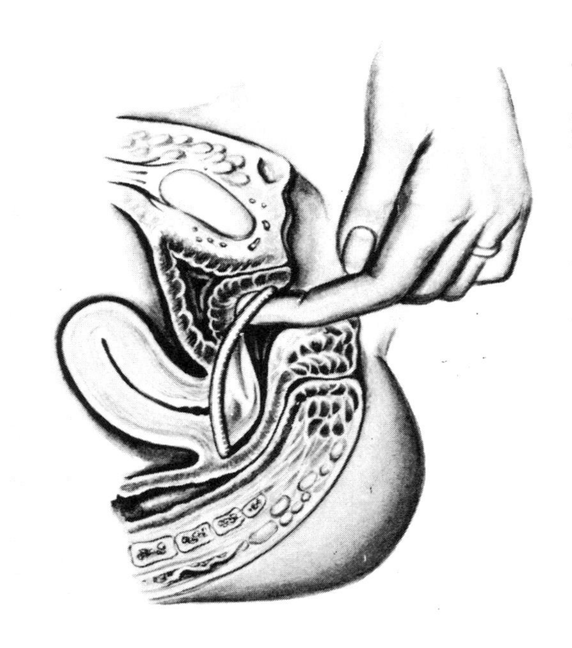

FIGURE 7-21. To remove the diaphragm, hook the outer rim of the diaphragm with the index finger and pull down and out. (From a manual for insertion of the Koromex® diaphragm, prepared by the Holland-Rantos Company, Inc., Piscataway, NJ, with permission.)

Problem: The patient would like to know if the diaphragm will disturb her partner.

Answer: Certainly not. If it does, the diaphragm has not been placed properly or was not fitted properly.

Problem: Can the diaphragm fall out?

Answer: Yes. Even the most properly fitted diaphragm can be dislodged during coitus, especially in the female superior position. Moreover, expansion of the vaginal canal during coitus will contribute to the dislodgement of an ill-fitted diaphragm. For this reason the diaphragm should be checked following coitus.

The major disadvantage of this method is its cumbersomeness. For this reason it is frequently not used, especially by young unmarried females with "irregular" sexual patterns. The diaphragm may not always be available when needed and the patient is not prepared. The patient all too frequently relies on her concept of the rhythm method in deciding when to use the diaphragm. There may also be a significant degree of anxiety as to the occurrence of the next menses.

Summary

Effectiveness: highly effective when used with a spermicide.

Advantages: initial medical instruction is required and the patient must learn how to use properly.

Disadvantages: cannot be used by certain women because of uterine variation; causes precoital interference; messy; privacy needed for insertion.

Risk of failure: variable but relatively low.

Alleged medical risk: none.

Oral Contraception

The birth control pill, introduced by Doctors John Rock, Gregory Pincus and Celso-Ramón García in 1956, is the most commonly used contraceptive in the United States. With some 10 million users in the United States and over 50 million users worldwide it has been readily accepted because of its high level of effectiveness and its ease of usage. The latter results in the dissociation of the use of the method and the sexual act.

The highest rate of usage among fecund women is in the Netherlands (37 percent in 1973), while total oral contraceptive usage is highest in China, where they are distributed free by "barefoot doctors," pharmacies and dispensaries. With the exception of Japan, India and the Union of Soviet Socialist Republics they are readily available throughout the world (Table 7-10). In addition to these facts, "the pill" is very popular among Roman Catholics and provides better sexual relationships: the coital frequency for white Catholic females using the pill is 45 percent higher than those using other methods.[24]

Despite this high level of acceptance, there has been much controversy surrounding its use, much of which is unjustified. We shall try to clarify some of the mysteries surrounding the pill so the physician will feel more informed about prescribing this method and better able to answer the patient's questions about it.

The effect of the pill on increasing coital frequency has been noted and may result from decreased anxiety about pregnancy, improved

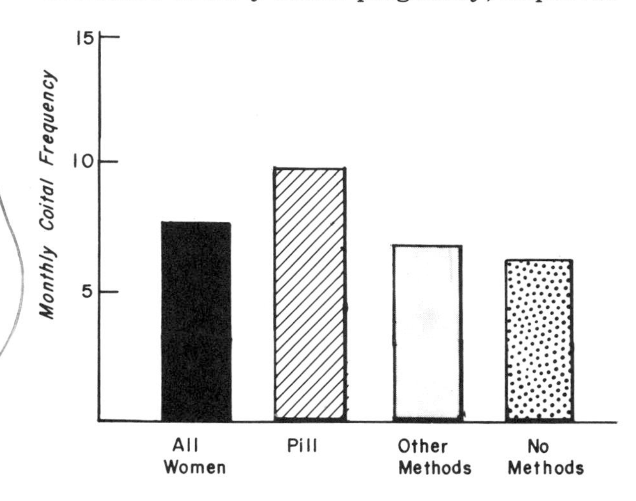

FIGURE 7-22. Mean monthly coital frequencies for women using the pill, other methods, or no method. (Data from Westoff, C. F., Bumpass, L., and Ryder, N. B.: *Oral contraception, coital frequency, and the time required to conceive.* Soc. Biol. 16:4, 1969.)

Table 7-10. Oral contraceptive sale—percent of total world sales.

	1968	*1969*	*1970*	*1971*	*1972*
Canada and U.S.	57.3	48.2	47.5	44.2	42.7
Central and South America	13.5	17.6	15.4	21.3	21.5
Europe	33.0	35.7	38.6	41.3	43.2
Japan and Philippines	0.8	1.4	1.1	1.0	1.0
Australia and New Zealand	——	9.0	8.4	8.2	10.9
World total	152.8	130.4	185.1	219.6	248.7

Data from *Oral Contraceptives,* Population Report, Series A, No. 1, April 1974.

sexual spontaneity resulting from the elimination of mechanical factors, and questionable endocrine interaction on libido. (Libido alterations are generally enhanced, but occasionally diminished.) A comparison of coital frequencies with various methods is shown in Figure 7-22.

Commonly used steroids in oral contraceptives are indicated in Figure 7-23 while types of oral contraceptives currently available and their breakdown are listed in Table 7-11.

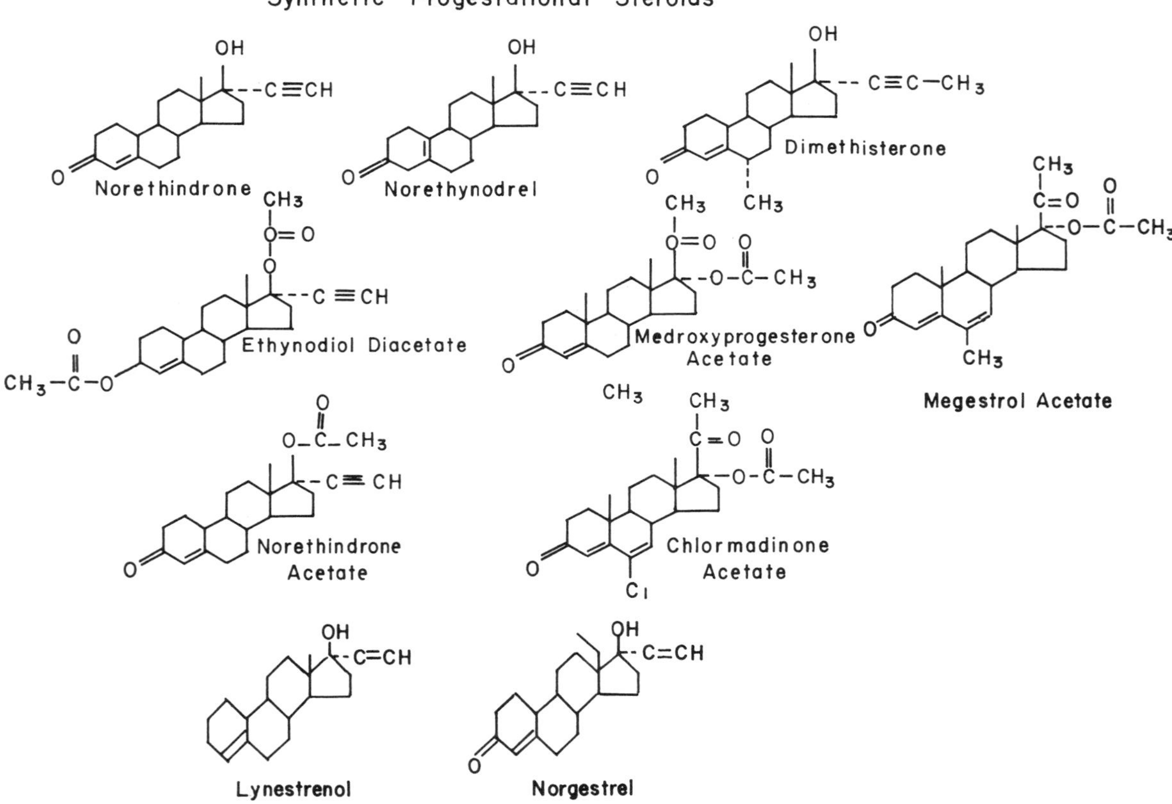

FIGURE 7-23. Commonly used steroids in oral contraceptives.

Table 7-11. Types of oral contraceptives currently available in the U.S.

Year introduced	*Product*	*Manufacturer*	*Type*	*Progestin*	*Estrogen*
1961	Enovid	Searle	Comb.	10 mg norethynodrel	150 mcg mestranol
1962	Enovid	Searle	Comb.	5 mg norethynodrel	75 mcg mestranol
1963	Ortho-Novum	Ortho	Comb.	10 mg norethindrone	60 mcg mestranol
	Ortho-Novum	Ortho	Comb.	2 mg norethindrone	100 mcg mestranol
1964	Enovid-E	Searle	Comb.	25 mg norethynodrel	100 mcg mestranol
	Norlestrin	Parke-Davis	Comb.	25 mg norethindrone acetate	50 mcg ethinyl estradiol
	Norinyl	Syntex	Comb.	2 mg norethindrone	100 mcg mestranol
1965	Provest*	Upjohn	Comb.	10 mg medroxypro-gesterone	50 mcg ethinyl estradiol
	C-Quens*	Lilly	Comb.	2 mg chlormadinone acetate	80 mcg mestranol
	Oracon†	Mead Johnson	Seq.	25 mg dimethisterone	100 mcg ethinyl estradiol
1966	Ovulen	Searle	Comb.	1 mg ethynodiol diace-tate	100 mcg mestranol
	Ortho-Novum SQ†	Ortho	Seq.	2 mg norethindrone	80 mcg mestranol
1967	Norinyl 1+50	Syntex	Comb.	1 mg norethindrone	50 mcg mestranol
	Ortho-Novum 1+50	Ortho	Comb.	1 mg norethindrone	50 mcg mestranol
	Norlestrin	Parke-Davis	Comb.	1 mg norethindrone acetate	50 mcg ethinyl estradiol
	Norquen†	Syntex	Seq.	2 mg norethindrone	80 mcg mestranol
1968	Ovral	Wyeth	Comb.	0.5 mg norgestrel	50 mcg ethinyl estradiol
	Ortho-Novum 1+50	Ortho	Comb.	1 mg norethindrone	80 mcg mestranol
	Norinyl 1+80	Syntex	Comb.	1 mg norethindrone	80 mcg mestranol
1970	Demulen	Searle	Comb.	1 mg ethynodiol diacetate	50 mcg ethinyl estradiol
1973	Micronor	Ortho	Prog.	0.35 mg norethindrone	
	Nor-Q.D.	Syntex	Prog.	0.35 mg norethindrone	
	Ovrette	Wyeth	Prog.	0.075 mg norgestrel	
	Loestrin 1.5+30	Parke-Davis	Comb.	1.5 mg norethindrone acetate	30 mcg ethinyl estradiol
	Loestrin 1+20	Parke-Davis	Comb.	1 mg norethindrone acetate	20 mcg ethinyl estradiol
1974	Zorane 1+20	Lederle	Comb.	1 mg norethindrone acetate	20 mcg ethinyl estradiol
	Zorane 1+50	Lederle	Comb.	1 mg norethindrone acetate	50 mcg ethinyl estradiol
	Zorane 1.5+30	Lederle	Comb.	1.5 mg norethindrone acetate	30 mcg ethinyl estradiol
1975	Modicon†	Ortho	Comb.	0.5 mg norethindrone	35 mcg ethinyl estradiol
	Brevicon	Syntex	Comb.	0.5 mg norethindrone	35 mcg ethinyl estradiol
	Lo/Ovral	Wyeth	Comb.	0.3 mg norgestrel	30 mcg ethinyl estradiol
1976	Ovcon-50	Mead Johnson	Comb.	1.0 mg norethindrone	50 mcg ethinyl estradiol
	Ovcon-35	Mead Johnson	Comb.	0.4 mg norethindrone	35 mcg ethinyl estradiol

*discontinued 1970 †discontinued 1976

Proposed Mechanisms of Action

The pill probably functions to prevent fertility in more than one way. Some of the proposed mechanisms of action (Fig. 7-24) are listed below:

1. Hypothalamic and pituitary inhibition resulting in inhibition of ovulation.
2. Change in fallopian tube motility and secretion.
3. Endometrial changes modifying nidation and possibly capacitation—gland regres-

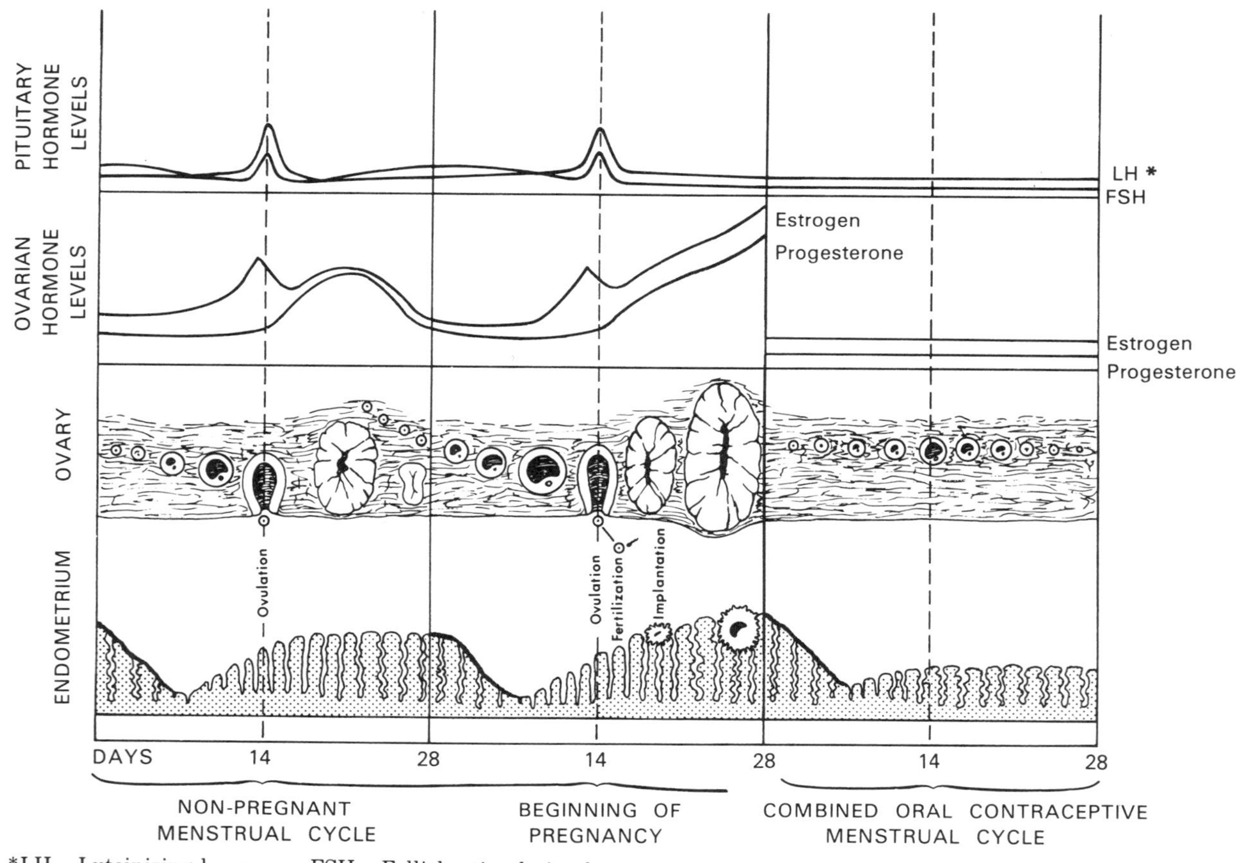

FIGURE 7-24. A diagramatic comparison showing changes which occur in the endometrium, ovary and in plasma hormone levels during the nonpregnant menstrual cycle, the beginning of pregnancy and the oral contraceptive menstrual cycle. (From Piotrow, P. T.: *Advantages of orals outweigh disadvantages*. Population Reports, Series A, No. 2, March 1975, p. 33, with permission.)

sion, predecidual changes, decreased blastokinin.

4. Hostility of cervical mucus to spermatozoa.

Physiologic Effects on the Reproductive System

The following are changes in organs or tissue of the reproductive system, resulting from the use of oral contraceptives. (In this and the following pages, the discussions will pertain to the more commonly used combination oral contraceptive. A description of the sometimes subtle differences between this and the sequential pill, which recently was taken off the market, would tend more to confuse than to educate.)

Vagina: low maturation index.

Problem: Patient who has been on the pill presents with the complaint of vaginal pruritus. What is her most likely problem?

Answer: Although any of the common causes of vaginitis can be found in women on the pill and should be investigated, the vaginal epithelium of the patient on the pill is often more susceptible to candidiasis. The reason for this is unknown.

Cervix: diminution, thickening and increased cellularity of the cervical mucus.

Problem: The patient calls your office and is concerned about a story she read in her women's magazine. She read that the pill causes cancer-like changes in her cervix.

Answer: The patient should be reassured. The pill, especially the progesterone component thereof, may be responsible for a benign readily-reversible atypical endocervical hyperplasia in 2 to 3 percent of oral contraceptive users. The cervix becomes lined with high columnar epithelial cells with abundant intracellular secretory products and a papillary convolution. Preliminary studies implicate a causal relationship but it is not considered hazardous.[25]

Endometrium: There is a prompt progression from proliferative endometrium through secretory endometrium to secretory exhaustion with occasional variable degree of decidual changes.

Problem: Patient asks, "What will happen to my menstrual flow?"

Answer: Because of this endometrial regression, pill users most frequently note a diminished flow.

Problem: Patient is concerned because she missed her period.

Answer: Two percent of the cycles on the pill are notable for the absence of withdrawal flow. Assuming the patient has not missed any pills, she should be reassured and reexamined if she misses two consecutive cycles and pregnancy should be ruled out.

Problem: The patient is concerned about breakthrough bleeding.

Answer: The patient should be advised to proceed with respect to the amount of flow. If the flow is heavy she should stop the pill and start again in five days, as if she were starting a new cycle. If the bleeding is scanty she should continue with one pill daily. Alternative treatment has been suggested.

The patient has been advised to double-up on her pills, using a spare pack, until the bleeding stops. At this time she should continue with one pill daily. She may also use supplemental estrogen preparations such as conjugated estrogen 0.625 mgm. to control the bleeding. These regimens may be confusing and misleading. Persistence suggests she should be reexamined and perhaps started on a different pill. The lower the estrogen dose in the pill, the more frequent is the breakthrough bleeding. However, pathologic causes of bleeding, albeit rare, must be ruled out in cases of persistent bleeding.

Ovary: There is an arrest in follicular development with low excretory levels of estriol, estradiol, estrone and pregnanediol.

Gonadotropins: Absence of luteinizing hormone (LH) peak and a slight decrease in follicle-stimulation hormone (FSH).

Side Effects

Most side effects noted by the patient are minor; frequently the side effects are beneficial. Adverse side effects are depicted in Figure 7-25 and beneficial side effects in Figure 7-26. The relationship of many of these side effects with the pill is sometimes questionable; the conditions may be related to other factors (Figs. 7-27, 7-28 and 7-29).

The more common side effects noted by patients are the following:

Breasts. The patient may note a transient increase in breast size as well as occasional mastalgia. These generally will regress after several cycles.

Skin. Chloasma or hyperpigmentation is a leading cause of pill discontinuation. This generally regresses on discontinuation. Acne, although generally improved on the pill, is occasionally worsened. Boils, rash, and alopecia have been reported to occur but are quite rare.

Central Nervous System. Libido alterations have been noted; they are generally enhanced but occasionally diminished. Depression, headaches, visual disturbances and dizziness have been described in patients.

Genitourinary System. Nausea and vomiting and other early pregnancy symptoms are occasionally noted in early cycles and frequently abate with use.

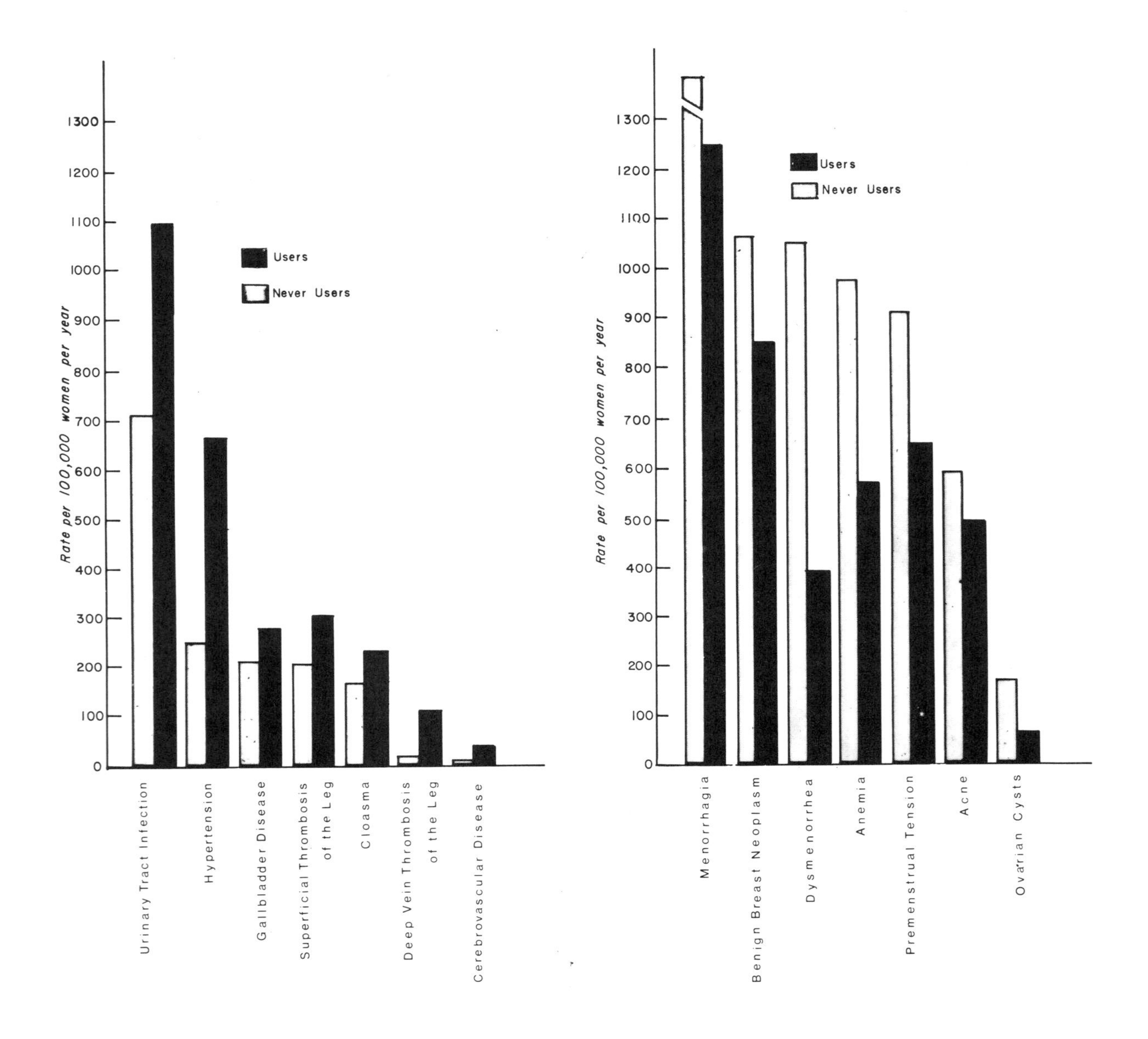

FIGURE 7-25. Adverse side effects of oral contraceptives in current users and never users in Britain from 1968 to 1971. These conditions occur more frequently among oral contraceptive users. (Adapted from Piotrow, P. T.: *Advantages of orals outweigh disadvantages.* Population Reports, Series A, No. 2, March 1975, p. 31.)

FIGURE 7-26. Beneficial side effects of oral contraceptives in current users and never users in Britain from 1968 to 1971. Note that pills have a protective effect against these conditions. (Adapted from Piotrow, P. T.: *Advantages of orals outweigh disadvantages.* Population Reports, Series A, No. 2, March 1975, p. 31.)

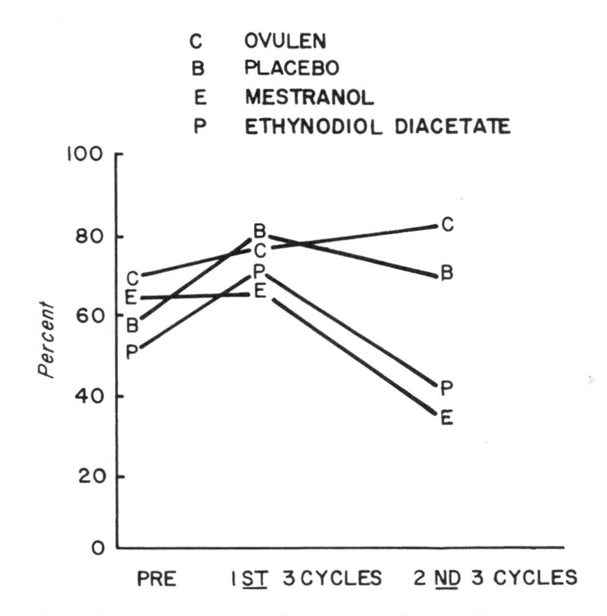

FIGURE 7-27. Total complaints in a double blind study. (Adapted from García, C.R.: The Second International Norgestrel Symposium, Royal College of Physicians, London, England, 1974. Excerpta Medica, p. 68.)

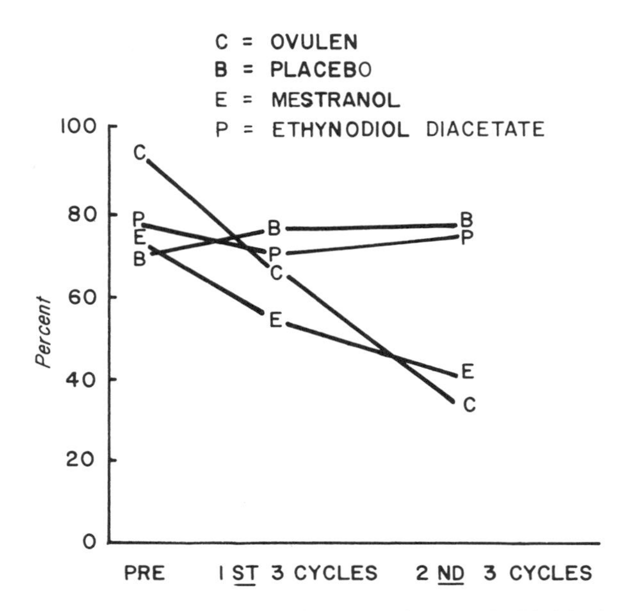

FIGURE 7-28. Incidence of dysmenorrhea in a double blind study. (Adapted from García, C.R.: The Second International Norgestrel Symposium, Royal College of Physicians, London, England, 1974. Excerpta Medica, p. 68.)

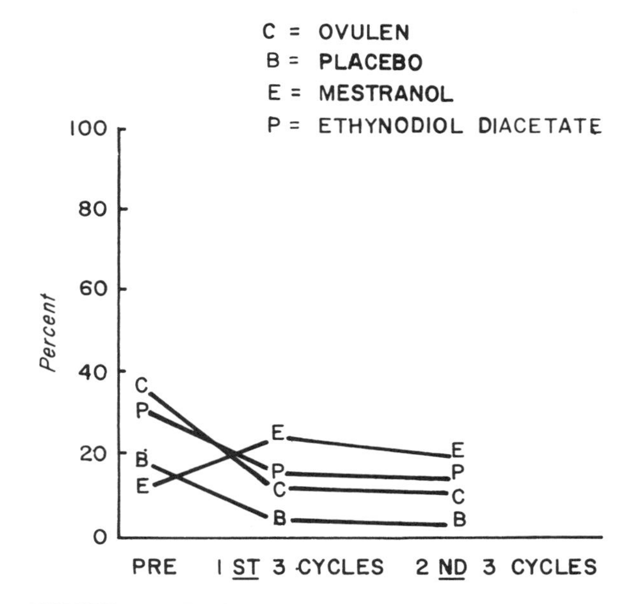

FIGURE 7-29. Incidence of breast tenderness in a double blind study. (Adapted from García, C.R.: The Second International Norgestrel Symposium, Royal College of Physicians, London, England, 1974. Excerpta Medica, p. 68.)

Problem: Patient calls and claims she has morning sickness.

Answer: Advise her to take pill in morning instead of evening. If side effects persist, switch to lower estrogen pill.

Problem: Patient is concerned about fluid retention and would like to know what to do about it.

Answer: Significant edema is quite unusual on the pill. If it should occur discontinuation of the pill is recommended.

Weight changes may occur but rarely more than five pounds gained or lost. An equal number of patients note a decrease in weight as those noting an increase.

Abdominal cramps and bloating have been noted by some.

Cardiovascular System. Easy bruisability (secondary to increased capillary fragility) and edema have infrequently been reported.

In summary, all of the above are not serious in nature, do not necessarily occur in every pa-

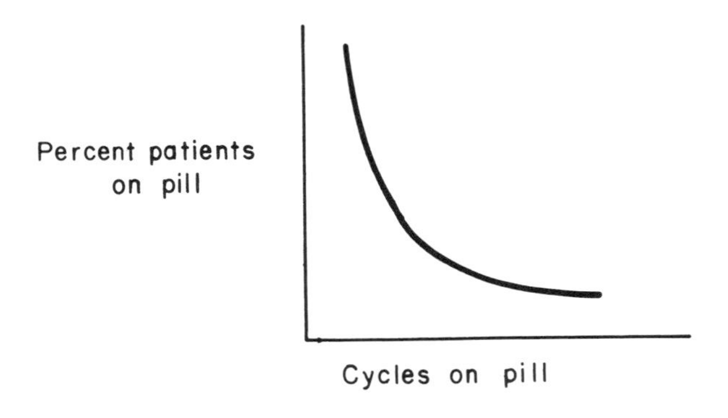

FIGURE 7-30. Occurrence of side effects of patients on the pill.

tient and most generally will diminish if not disappear entirely after several cycles of using the pill (Fig. 7-30). Care in describing the side effects will increase the patient's compliance. Many of the minor side effects, especially those associated with pseudopregnancy (nausea, vomiting, headaches, fluid retention, weight gain and mastalgia) will abate with usage.

Laboratory Alterations

The physiologic effects of the oral contraceptive can distort the results of numerous analytic tests, thereby either suggesting disease when there is none or obscuring disease when it is present (Table 7-12).

TABLE 7-12. Potential effects of oral contraceptives on the results of laboratory tests.

Determination	*Significant effect*			*Remarks*
	Slight	*Moderate*	*Marked*	
Albumin (serum)		Decreased		
Aldosterone (blood)	Increased	Increased		
Aldosterone excretion (urine)		Increased		
Alkaline phosphatase (serum)	Decreased	Increased		Increased in cholestasis
Alpha amino nitrogen (serum)		Decreased		Progestogenic effect
Alpha-1 antitrypsin (serum)		Increased		
Alpha-1 globulin (serum)		Increased		
Alpha-2 globulin (serum)			Increased	
Angiotensin I and II (serum)			Increased	
Angiotensinogen (serum)		Increased	Increased	
Antinuclear antibody test		Positive		
Antithrombin III (serum)			Decreased	
Ascorbic acid (plasma)		Decreased		
Ascorbic acid (in leukocytes)		Decreased		Not increased by supplemental ascorbic acid
Bilirubin (serum)		Increased		Reduced hepatic excretion; preexisting hyperbilirubinemia accentuated (Dubin-Johnson)
Sulfobromophthalein retention (serum)		Increased		Related to hepatic transfer mechanisms
Butanol extractable iodine (serum)		Increased		Result of increased thyroxine-binding globulin
Calcium (serum)	Decreased			
Calcium excretion (urine)	Decreased			
Cephalin flocculation (serum)			Increased	
Ceruloplasmin (serum)		Increased		
Cholesterol (serum)	Increased	Increased		Variable response to various preparations; decreased when therapy discontinued; no change in some reports
Coagulation factor II (plasma)	Increased			Usually unchanged
Coagulation factor VII (plasma)		Increased		Estrogenic effect
Coagulation factor VIII (plasma)	Increased			Rarely affected; change usually not statistically significant
Coagulation factor IX (plasma)	Increased			Usually unchanged
Coagulation factor X	Increased			Usually unchanged
Coagulation factor XII (plasma)	Increased			Usually unchanged
Copper (serum)			Increased	Increased synthesis of ceruloplasmin: aggravated by cholestasis

TABLE 7-12. (Cont'd)

Determination	*Significant effect*			*Remarks*
	Slight	*Moderate*	*Marked*	
Coproporphyrin (fecal and urine)			Increased	
Cortisol (blood and urine)		Increased		Estrogen causes decreased cortisol clearance; non-protein-bound cortisol increased; no increase with progestogen
Complement-reactive protein (serum)		Increased		By estrogen
Cryofibrinogen (plasma)		Increased		
Erythrocyte count (blood)	Decreased			Slight and proportional decrease in hemoglobin and red blood cell indices
Estradiol excretion (urine)		Decreased		
Estrogens, total (urine)			Increased	
Etiocholanone excretion (urine)	Decreased			
Euglobulin lysis (plasma)	Increased	Increased		Whole blood lysis and other parameters of fibrinolysis also increased
Fibrinogen (plasma)	Increased			Estrogens alone; usually normal
Formiminoglutamic acid excretion after histidine (urine)		Increased		
Folate (serum)			Decreased	Long-term use may cause megaloblastic anemia; poor absorption of polyglutamic folate
Follicle-stimulating hormone excretion (urine)			Decreased	Long-term use
Glucose tolerance (blood)	Decreased	Decreased		Varies with drug used; fasting blood glucose usually not elevated except in prediabetic patients; prednisone glucose tolerance most abnormal with mestranel alone; marked effect in prediabetic subjects; tends to return to normal during long-term therapy
Gonadotropin excretion (urine)		Decreased		
Growth hormone (serum)		Increased		During first year
Haptoglobin (serum)	Decreased			
Hematocrit (blood)		Increased		Progestogenic effect
Immunoglobulin A (serum)	Increased			
Immunoglobulin G (serum)	Increased			
Immunoglobulin M (serum)	Increased			
Insulin (serum)		Increased		Not increased above pretreatment level in overt diabetic patients
Iron (serum)		Increased		
Iron-binding capacity (serum)			Increased	
17-ketosteroid excretion (urine)		Decreased		Moderately increased in some cases
Lactate (blood)		Increased		
Lupus erythematosus cell preparation (blood)		Positive		May exacerbate preexisting systemic lupus
Leukocyte count (blood)		Increased		Hemophilic children
Luteinizing hormone (blood and urine)			Decreased	Effect additive with combination-type pill
Lipoproteins, alpha (serum)		Increased	Increased	
Lipoproteins, beta (serum)			Increased	
Lipoproteins, prebeta (serum)		Increased		
Lipoproteins, total (serum)			Increased	
Lymphocyte transformation (phytohemagglutinin)			Decreased	
Magnesium (serum and urine)		Decreased		
Nucleotidase (5-nucleotidase) (serum)	Increased			Related to cholestasis
17-hydroxycorticosteroid excretion (urine)		Decreased		
Partial thromboplastin time (plasma)	Increased			Usually no significant change
Protein-bound iodine (serum)		Increased		Occasionally markedly increased
Phospholipids, total (serum)			Increased	
Plasma volume		Increased		

TABLE 7-12. (Cont'd)

Determination	*Significant effect*			*Remarks*
	Slight	*Moderate*	*Marked*	
Plasmin (plasma)		Increased		
Plasminogen (plasma)		Increased		
Platelet adhesiveness (blood)	Increased			
Platelet aggregation (adenosine diphosphate) (blood)		Increased		Combination estrogen-progestogen
Platelet count (blood)	Increased			
Porphobilinogen excretion (urine)			Increased	
Pregnanediol excretion			Decreased	Pregnanetriol also decreased
Prothrombin time (plasma)	Decreased			Decreased response to oral anticoagulants
Prothrombin time (plasma)		Increased		If cholestasis occurs
Protoporphyrin (fecal)			Increased	
Pyruvate (blood)		Increased		More increased after glucose administration
Renin (serum)			Decreased	Plasma concentration decreased but renin activity increased; greater increase in hypertensive subjects
Erythrocyte sedimentation rate (blood)	Increased			
Serum glutamic oxalacetic transaminase, aspartate aminotransferase (serum)	Increased			Usually no significant change
Serum glutamic pyruvic transaminase, alanine aminotransferase (serum)	Increased			Usually no significant change
Sodium (serum)		Increased		
Testosterone (serum)		Increased		
Tetrahydrocortisone (urine)		Decreased		
Thyroid-binding globulin (serum)		Increased		
Thyroxine (serum)		Increased		
Total lipids (serum)			Increased	
Transferrin (serum)			Increased	Conflicting data
Triglycerides (serum)		Increased	Increased	By estrogen but not by some combinations, variable response to combinations, marked increase in preexisting hypertriglyceridemia
Triiodothyronine (serum)		Increased		Occasionally markedly decreased
Urobilinogen excretion (urine)		Decreased	Decreased	
Uroporphyrin excretion (urine)		Increased		
Vitamin A (plasma)			Increased	
Vitamin B_{12} (serum)		Decreased		Vitamin B_{12} binding capacity increased
Xanthuric acid excretion (urine)			Increased	Other tryptophan metabolites also increased, related to vitamin B_6 deficiency
Zinc (serum)		Decreased		

From Miale, J. B., and Kent, J. W.: *The effects of oral contraceptives on the results of laboratory tests.* Am. J. Obstet. Gynecol. 120:265, 1974.

Risks Involved

Having dealt with the various side effects, physiologic changes and laboratory alterations most commonly noted with the pill, one should consider the relative risks of taking the pill. The pill has been considered dangerous by some and totally innocuous by others. Although the truth probably lies somewhere between these two extremes, one should be aware of the extent of the risks in order to evaluate the feasibility of prescribing the pill for the individual patient.

In determining the relative risks of oral contraception, one must first take into consideration the risks of *not* practicing fertility control or the risks involved in using other methods (Table 7-13).

Of greater importance to the physician prescribing the pill is the difference in the age of the patient with respect to the risk as well as the risk compared to other hazards of life (Table 7-14).

Table 7-13. Mortality associated with pregnancy and fertility control in developed and developing countries. (Annual rates per 100,000 women aged 15 to 40)

Method	*Estimated method failure/HWY*	*No. of pregnancies from method failure*	*No. of deaths from pregnancy (25 per 100,000 births)*	*Estimated no. of deaths from method*	*Total deaths from pregnancy and method*
		DEVELOPED COUNTRIES			
No contraception	0.0	60,000	15	0	15
Oral contraception	1.0	1,000	<1	3	<4
IUDs	3.0	3,000	<1	1	<2
Condom & diaphragm	20.0	15,000	4	0	4
		DEVELOPING COUNTRIES			
No contraception	0.0	60,000	300	0	300
Oral contraception	2.0	2,000	10	3	13
IUDs	3.0	3,000	15	1	16
Condom & diaphragm	20.0	15,000	75	0	75

Data from Piotrow, P.: *Advantages of orals outweigh disadvantages.* Pop. Reports, Series A, No. 2, March 1975 p. 35.

Table 7-14. Comparative estimated risks and hazards of users and non-users or oral contraceptives.

	Ages in Years	
	26–44	35–44
E.A.D.R./10⁵ healthy married non-pregnant women from pulmonary embolism or cerebral thrombosis*		
Users of oral contraceptives	1.5	3.9
Non-users of oral contraceptives	0.2	0.5
*A.D.R. */10⁵ total female population from:*		
Cancer	13.7	70.1
Motor accidents	4.9	3.9
All causes	60.0	170.5
Death rate/10⁵ maternities from:		
Complications of pregnancy,	7.5	13.8
All risks of pregnancy, delivery and puerperium	22.8	57.6

Data from Inman, W. H. W., and Vessey, M. P.: *Investigation of deaths from pulmonary, coronary, and cerebral thrombosis and embolism in women of child-bearing age.* Br. Med. J. 2:193, 1968.

*Key:
E.A.D.R.–estimated annual death rate
A.D.R.–actual annual death rate

Therefore one can conclude that (1) although the pill is not innocuous, the relative risk is very low; (2) the pill is safer than pregnancy or any other type of contraceptive with a decreased effectiveness which would result in more method failures; and (3) the pill is safer in the younger patient (below age 35).

Thromboembolic Disease

There have probably been fewer more highly debated medical items confronting the general public than the alleged association between the pill and blood clots. The major result of the Congressional hearing of 1972 was an increase in the pregnancy rate among women stopping the pill out of fear of developing a clotting disorder. The statistics, based mainly on retrospective study, can be criticized. Moreover, the overall risk to the patient, as we shall see, is very small.

Etiology: It has been suggested that the

thromboembolic phenomenon is related to the change in blood coagulation factors in women on the pill. These changes include

1. platelets—increase in count in most long term users
2. rise in fibrinogen and factors 2, 8, 9 and 12 and a decrease in clotting inhibitor antithrombin III
3. increase in fibrinolysis
4. increase in triglycerides and pre-beta-lipoproteins
5. adventitial changes in blood vessels
6. dilatation of vascular bed

Note that (1) in vitro studies have failed to demonstrate coagulation; (2) there are great individual variations in these factors; (3) bleeding, clotting and prothrombin times are not altered; and (4) vascular changes 5 and 6 have been questioned.

Incidence. It has been estimated that at most one woman in two thousand on the pill may suffer a blood clotting disorder severe enough to require hospitalization. The death rate from thromboembolism, pulmonary embolism and cerebrovascular accidents among these women is reported to be much less (Table 7-15).

Table 7-15. Estimated death rate from thromboembolism among women of reproductive age.

under 35	1/500,000
under 35 (on pill)	1/66,000
over 35	1/200,000
over 35 (on pill)	1/25,000

From Frick, K. E.: *Medroxyprogesterone acetate injectable contraceptive.* Federal Register 38:27941, 1973.

Evidence. Controlled retrospective studies in England and the United States[26,27,28,29] have suggested an etiologic relation of the pill and thromboembolic disease. They have found

1. The incidence of idiopathic thromboembolic disease appears to be several times higher among females taking the oral contraceptives. There is (a) a 50 percent increase in the risk of superficial thrombosis in the leg (from 200 to 300 cases per 100,000 women per year); (b) a five- to sixfold increase in the risk of deep vein thrombosis (from 20 to 110/100,000); and (c) a fourfold increase in the risk of stroke (from 10 to 40/100,000 women yearly).

2. The incidence was reported to be directly associated with the estrogen content of the tablet. However, by elimination of the highest dose preparation (150 gamma) which no longer is used, no statistically significant difference could be demonstrated between the 50, 80 or 100 gamma pill. These calculations were based on sales in Europe at that time.

3. The risk was found to be less in women with blood group O.

4. The risks are readily reversible with discontinuation of the pill.

5. There is no association with the duration of pill use.

To further confuse (or clarify) the issue there has been a significant increase in the incidence of idiopathic thromboembolic disease in Western culture since 1958, occurring in both sexes and among women not using the oral contraceptive (Fig. 7-31). This has not been noted in Eastern countries where both the use of tobacco and the incidence of obesity is much less.

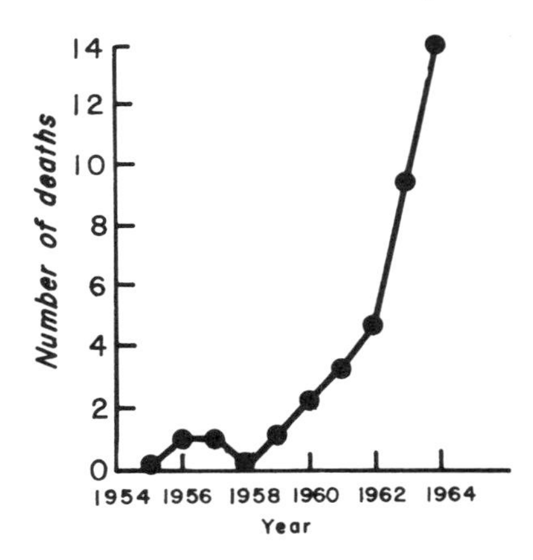

FIGURE 7-31. Deaths from idiopathic pulmonary embolism in adults under age 40 per 18,300 in patient population in London, England. No Pill users were among the fatalities. (Adapted from Goldzieher, W.W.: *An assessment of the hazards and metabolic alterations attributed to oral contraceptives.* Contraception 1:409, 1970.)

Following the announcement in Great Britain at the end of 1969 regarding estrogen dosage in the oral contraceptive pill and the associated risk of thrombosis and the suggestion that women be switched to low estrogen dose pills, there was a 15 percent discontinuation rate of the pill. The consequence was the first increase in the British birth rate since 1964 and marked increase in the number of abortions.

Conclusion. The possibility of an increase in the incidence of thromboembolic disease occurring in patients on the pill does exist. However, if it is real, the risk is very low (three deaths yearly per 100,000 users attributed to oral contraceptives), especially when compared with other hazards of life which are avoidable (i.e., smoking, drinking, other contraceptives and pregnancy itself). Moreover, the risks involved could possibly be further minimized by good patient selection (e.g., avoid older patients, those with obesity and varicosities).

Problem: A patient on the pill is complaining of frequent severe headaches.

Answer: Stop the pill, observe the patient and reevaluate. Of those reported cases of cerebrovascular accidents among healthy young women on the pill, the most frequent warning was a headache.[30]

Problem: A patient comes to you desiring the pill. She has had an episode of thrombophlebitis.

Answer: Be prudent and suggest another method. This rule should also be applied to other high risk patients, i.e. elderly females, women with varicose veins and very obese females.

Problem: A patient on the pill requires surgery. What should be done?

Answer: On the basis of retrospective studies suggesting an increased risk of post-surgery thromboembolic complications in oral contraceptive users, the U. S. Food and Drug Administration (FDA) recommends stopping oral contraceptives two weeks prior to, and not resuming them until at least two weeks after, surgery.[31,32]

Hypertension

Prospective studies have shown a slight but significant rise in systolic blood pressure (mean 6.6 mm/Hg) in most women on the pill, albeit they remained primarily in the range of a normotensive. Discontinuation usually results in a return to pretreatment levels within three months. Rare occurrences of hypertensive crises developing in women on the pill have been reported. Less than 5 percent of patients will develop clinical hypertension.[33,34,35] The hypertension is generally ascribed to the estrogen in the pill and is reputed to be more prone to occur in women with a previous history of hypertension. It is confusing that postmenopausal women, treated with estrogen, often have an improvement in their vascular status.

Suggested Mechanism. The increase in the blood pressure has been related to an alteration in the renin angiotensin system. Increasing levels of estrogen are associated with an increase in the hepatic production of the alpha globulin angiotensinogen, a renin substrate. There is also an increase in renin activity as well as angiotensin II with a decrease in the concentration of renin. A defect in the feedback control mechanism is postulated. A unanimity on this has not been achieved.

Problem: Should a patient discontinue the pill if there has been an abnormal rise in blood pressure?

Answer: Yes. The patient should be evaluated if this persists. Another form of fertility control should be offered. A trial with a different pill may be attempted if there are no medical contraindications.

Problem: Should I prescribe the pill to a patient with mild essential hypertension?

Answer: The safety of oral contraceptive administration to these women and the lack of effect on their blood pressure has been shown. In some cases, the blood pressure actually decreased while on the pill. It is prudent to watch these patients carefully nonetheless.[36]

In view of the occurrence of hypertension on the pill, all patients in whom contraception has been initiated should have their blood pressure checked at least once from one to three months after beginning therapy and at least twice a year thereafter.

Myocardial Infarction

An association between myocardial infarction and oral contraceptives has been suggested by two retrospective epidemiologic studies in the United Kingdom. A synergism between oral contraception and several risk factors was noted. These included obesity, age, hypertension, preeclamptic toxemia, hypercholesterolemia and cigarette smoking. The risks in women under 40 of fatal and nonfatal myocardial infarction attributable to oral contraception is nonetheless about 3.5 per 100,000 users per year. In those women with no other risk factor identified the risk is nonexistent.[37,38,39]

Problem: At what age should my patient stop taking the pill?

Answer: There is no good answer to that question. Although evidence suggests that older patients are at greater risk for pill-related complications, they are also at greater risk for pregnancy-related complications. Evaluate the patient and her situation. Consider the alternatives (i.e., sterilization or mechanical fertility regulation). The F.D.A. cautions against the use of the pill in women over 40.

Cancer

No association has been shown between the combination oral contraceptive pill and cancer of the breast or cervix. Neither has there been a recent change in the incidence of cancer of the breast,[40] cervix or endometrium (Fig. 7-32). Moreover, animal studies have failed to demonstrate an increase in the frequency of animal carcinomas with the use of combined estrogens and progestins as found in the combined pill.

In fact, the pill may have a beneficial effect on atypia of the endometrium. Also, a decrease in benign breast disease and benign ovarian neoplasm of the ovary has been reported. The pill does not adversely affect the PAP smear.

Problem: The patient requesting the pill informs you that her mother had breast cancer. Can she still take the pill?

Answer: Yes. Whereas the pill is contraindicated in patients with preexisting breast cancer, there is no contraindication to prescribing the pill to this patient.

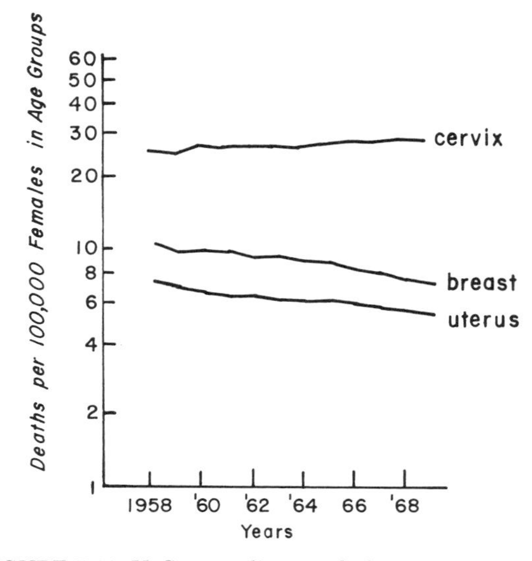

FIGURE 7-32. U. S. mortality trends for cancer of the cervix, breast and endometrium. (Adapted from Vokaer, R.: *Oral contraception and cancer of the uterus and breast.* In Persianinov, L. S., and Chervakova, T. V. (eds.): *Recent Progress in Obstetrics and Gynecology.* Excerpta Medica, Amsterdam, 1974.)

Liver

Abnormalities in excretory function (prolonged bromsulphalein excretion), enzyme concentrations and intracellular ultrastructure have been noted. They are innocuous and readily reversible. Cholestatic jaundice will occur in 1/10,000 users. There is a twofold increase in the incidence of cholelithiasis and/or cholecystitis in pill users (reported as 68 to 158 cases/100,000 users) perhaps related to the increase in the saturation of gallbladder bile with cholesterol, the increase in serum triglycerides or to a decrease in cholesterol excretion. This condition seems to be related to the amount of progestin and the duration of use.[41,42,43,44]

Problem: Should a patient with a history of recent hepatitis be on the pill?

Answer: The patient should first demonstrate normal liver function. She should be followed with liver function tests thereafter. The same is prudent for patients with infectious mononucleosis. The pill should not be given to females with a history of cholestatic jaundice of pregnancy or in those with chronic defects in hepatic excretory function (e.g., Dubin-Johnson syndrome). Patients with a history of pruritus gravidarum probably are poor candidates for the pill. Recently, over 30 cases of rare liver tumors, generally benign, have been reported in women on the pill, but no cause and effect relationship has been documented.[45,46] The adenoma cells differed very little from normal hepatocytes. Both long and short term users were included although in one study the relative risk was higher with prolonged use. Presenting symptoms were sudden severe abdominal pain or shock resulting from rupture and hemorrhage of the liver tumor. Suspicion of hepatomegaly should be investigated with a liver scan.[47]

Diabetes

Early and persistent alteration in carbohydrate metabolism has been noted in women on the pill. These include an abnormal glucose tolerance test and increase in plasma insulin. These changes are readily reversible.

Problem. The patient would like to know if she stands a greater chance of developing diabetes while on the pill.

Answer. No overt diabetes has developed in pill users studied for over 15 years.[48]

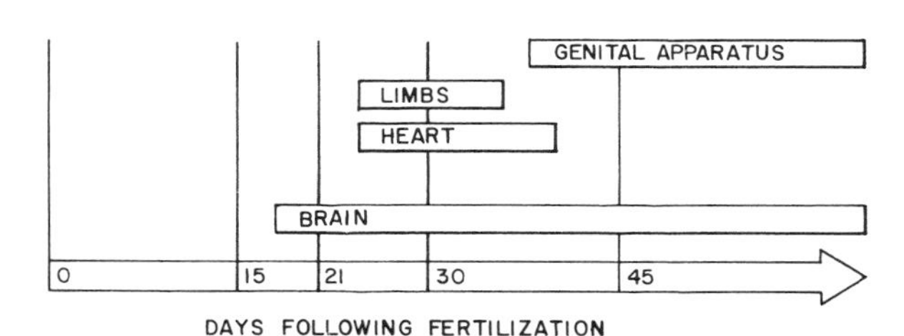

FIGURE 7-33. Development of various embryonic systems at which time drugs may act.

Problem. Should the pill be prescribed for the insulin dependent diabetic?

Answer. Yes. Some of these patients may require an increase in their insulin requirements while on the pill. Occasionally a decrease in insulin requirement has been observed. Patients who are not insulin dependent and who have a history or suspicion of gestational diabetes should be cautioned and watched carefully.

Teratogenicity

Teratogenic effects of a particular drug can act at different levels (Fig. 7-33). These can affect the embryo itself, the gamete of the embryo or the gamete of the parent prior to fertilization. Effects could result in (1) embryo, fetal or neonatal death; (2) minor or major fetal malformations with survival; or (3) damage to reproductive process, such as future fertility impairment or chromosomal defects in future offspring.

The following have been noted to be associated with hormonal steroids. Certain questions still remain unanswered.

1. Masculinizing effects have been noted on the external genitalia of female fetuses born to mothers who had received elevated doses of some progestins during early pregnancy.
2. An increase was reported of certain chromosomal abnormalities in spontaneously aborted embryos from women who conceived within six months following the discontinuation of oral contraceptives.[49]
3. However, the incidence of abnormalities in neonates born to previous oral contraceptive users was not different from that of a matched group of non-users.[50]
4. Recent studies[51,52,53] have suggested a relationship between exposure to sex steroids during pregnancy (especially hormonal pregnancy tests) and certain limb reduction defects as well as other (VACTERL) vertebral, anal, cardiac, tracheal, esophageal, renal, and limb anomalies. These defects, if accurate, would occur in seven births per 10,000 pill failures in one million users. Therefore as a precaution, hormonal pregnancy tests should not

be used. Since effects of the birth control pills are dissipated in less than 48 hours in blood detection the advice to use mechanical contraception for two to three cycles following pill discontinuation before conception is attempted is probably excessive.

5. Chromosome breakage in lymphocytes and fibroblasts from women on the pill as compared to controls has not been shown to occur.[54]

Postpill Amenorrhea

Most patients will ovulate 4 to 8 weeks after stopping the pill. Approximately one patient in 500 will wait longer than six months for her first period. Fifteen percent of these patients may have an associated galactorrhea.

Problem: As a physician, can I predict which patients will be risks for developing this problem?

Answer: Not really. Patients with preexisting hypothalamic-pituitary-ovarian dysfunction would be expected to have problems regardless of oral contraceptive use. To avoid confusion, another type of fertility control probably should be offered to these women, or the responsibility should be shared with the patient. Moreover, there probably is little justification in regulating the cycles of an oligomenorrheic patient unless endometrial hyperplasia is being treated.

Problem: Should I stop the pill and allow the patient to ovulate spontaneously every so often to prevent this from occurring?

Answer: No. There is no correlation with the duration of therapy and the occurrence of postpill amenorrhea. Moreover, periodic interruption of oral contraception is not justified—pregnancy is at risk, and the patient is inconvenienced without benefit.[55]

Problem: How do I treat this entity?

Answer: Over ninety percent of these patients will eventually resume normal menses within six to twelve months after onset. Treatment with clomiphene citrate and menotropins has been effective in establishing ovulation in the majority of the rest of the patients. Return of normal ovulatory cycles on discontinuation should not be guaranteed. Moreover, be aware of the possible coincident occurrence of intracranial lesions. Certainly in all patients with galactorrhea-amenorrhea, tomography is mandatory.[56]

Pregnancy

A potential risk of oral contraceptive usage on future reproduction has not been demonstrated.

Problem: The nulligravida patient using the pill to delay a desired pregnancy would like to know if the pill will reduce her chances of subsequent pregnancy.

Answer: There is no difference in the time required to conceive between women who have used the pill and those using other forms of fertility regulation following cessation of that method. Eighty-five percent of pill users and 87 percent of nonpill users were pregnant by the end of six months with a mean waiting time of 2.3 months for the former and 2.1 months for the latter[57] (Fig. 7-34).

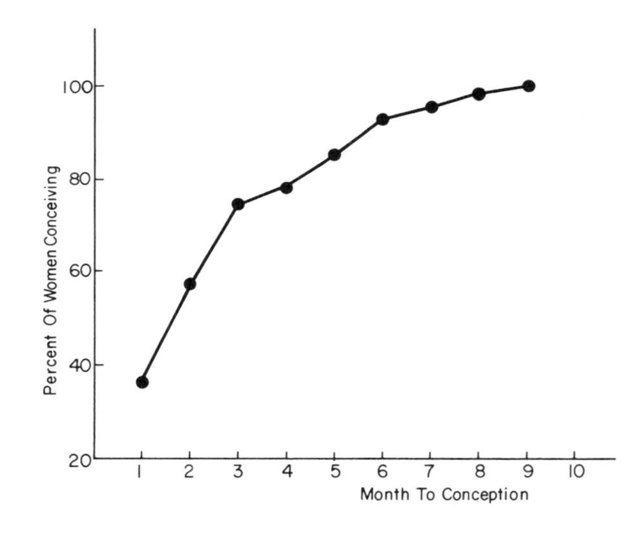

FIGURE 7-34. Percent of women who conceived after oral contraception within two years or less. (Modified from Goldzieher, J. W., and Hines, D. C.: *Large-scale study of an oral contraceptive.* Fertil. Steril. 19:841, 1968.)

Skeletal Maturation

The use of the pill by the very young adolescent presents a variety of problems. Among these are the effects of the pill on skeletal maturation.

Problem: A 13-year old sexually-active growing female is brought in by her mother for the pill. Will this stunt her growth?

Answer: To be effective as a deterrent to excess growth, high doses of estrogen must be started prepuberally. The use of estrogens in doses exceedingly larger than in the pill is not uniformly successful in inducing epiphyseal closure. The amount of estrogen produced during the normal menstrual cycle is greater than the exogenously administered estrogen during oral contraception usage.

Vitamin Deficiency

An increase in the plasma concentration of vitamin A and decreases in vitamins B_6, B_{12}, folic acid and C have been reported. These changes appear to be of no clinical significance, although disturbances of vitamin B_6 and tryptophan metabolism have been postulated to be associated with mental depression.[58]

Summary

Current opinion regarding the side effects of oral contraceptives are expressed as broad generalizations regarding a spectrum of numerous progestagens and principally two oral estrogens, i.e. ethinylestradiol and the 3-methyl ether of ethinylestradiol. Despite specific differences in the individual steroids, including dosages, more often all are grouped and discussed in terms of "the pill." These conclusions are derived from clinical impressions or projected principally from animal studies. Animal data is transposed liberally since human data often is unavailable. Retrospective studies and prospective studies with matched controls have been accepted to define the specific or comparative effects of the various combinations and even of the individual components. Few exacting comparative human pharmacologic studies are available. Double blind random assignment studies of the various oral contraceptives are difficult to achieve and even more so to interpret. It is obvious that blind random control or placebo assignment without the subject using another contraceptive of equal effectiveness is ethically unacceptable. The numerous contributing patient factors and physician contributions affecting the detection and reporting of side effects are legion. Moreover, while from clinical experience it may be feasible to deduce the *relative* incidence of side effects that may occur with great frequency, it becomes exceedingly difficult and perhaps impossible to do this when the incidence is exceedingly small. Table 7-16 indicates the enormity of the population sampling required to be able to detect a twofold increase when the frequency of incidence is exceedingly low. Such drug-induced actions with an annual incidence of 0.3 per 10,000 would require follow-up on 600,000 persons for one year or 80,000 for five years in both the treated as well as the control group in order to detect a twofold increase. Obvious logistical and other difficulties preclude being able to accomplish this. On the other hand, an effect with an annual incidence of 30 per 10,000 would merely require 600 persons studied for one year or 100 for five years. Such a sampling becomes logistically more feasible.

Table 7-16. Detection of twofold increase of side effects with number of patients followed.

Annual incidence (rate per 10,000)	*No. followed 1 year*	*No. followed 5 years*
0.3	600,000	80,000
3.1	60,000	7,500
20.0	9,000	1,200
30.0	600	100

Data from Siegel, D., and Corfman, P.: *Epidemiological problems with studies of the safety of oral contraceptives.* J.A.M.A. 203:950, 1968.

Unfortunately most of the serious hazards that have been viewed with concern in association with oral contraceptive usage fall within the lower frequency level of incidence. Accordingly the difficulty in being able to define a precise causal effect assessment becomes obvious. Such is the fundamental basis of much of the confusion that exists regarding the potential occurrence of hazardous effects with oral contraceptive usage. It must also be stressed that retrospective studies can only support that an association *may* be possible. Repeated confirmatory retrospective reviews do not strengthen the hypothesis. The very fundamental premise of scientific reasoning that relies on association and causality can be a weakness since association is not always synonymous with causality. Further proof with more objective measures supported through demonstration of mechanisms of action are needed.

An analogous situation relates to infectious disease. So often it is difficult to seek out such details and there follows a compulsion to take a short cut by neglecting the first two of the four Koch postulates which should be fulfilled in definitively establishing a causal effect relationship. Omitting the first two of the Koch postulates from the complete chain of events presents the potential of risk of an erroneous conclusion. The improper conclusion has caused many a missed diagnosis.

While this means admitting an inability to definitively assign a causal effect relationship regarding the oral contraceptives, one may be accused of nitpicking by not accepting a moderate probability of causal effect relationship. In making clinical decision nonetheless one should also be aware of the reality of all the implications. Once a regulatory restriction is issued or even suggested, the population exposed to the agent is less random and the incidence is potentially altered, further adding to the confusion. Moreover from a medicolegal standpoint far too much emphasis is placed on the validity of the retrospective studies and/or the conclusions from the inadequate prospective reviews. Certainly such conclusions do *not* warrant the inordinate settlements that have been rendered and which cannot but becloud the judgment of prescribing physicians and even more so of the third party carriers.

The public and the legal profession have been much too prone to overinterpret the well intentioned but rather inconclusive guidelines made in good faith by the scientific medical community. Such actions deter or even deny our citizens the use of the approaches which might be best in certain individual circumstances. Some of these statements which have been issued are misleading and may lead large numbers of women to abandon oral contraceptives in favor of less reliable methods, thus exposing them to the very real risks, both physical and emotional, of undesired pregnancy. It must be emphasized that irresponsibility cannot be eliminated in toto, thus no matter how many precautions are taken total safety will never be achieved. Nonetheless careful management requires diligent and interested supervision to detect possible eventualities whether or not they are causal effect related.

Fertility control is not the only benefit derived from oral contraceptive therapy. There also are therapeutic advantages for the following conditions:

Anemia—the lower incidence of iron deficiency anemia in pill users is attributed to decreased menstrual blood loss.

Menorrhagia—pathologic causes should first be ruled out.

Dysmenorrhea—majority of patients will find relief on the pill.

Premenstrual tension—relief in premenstrual anxiety, depression, lethargy and irritability often is noted by women on oral contraceptives.

Endometriosis—symptomatic relief can be achieved in some patients while on the pill.

Acne—frequently improved on pill.

Office Management

Although examination by a physician is advisable, it is not always possible. Therefore, the risks in withholding contraception for this reason must be evaluated. For women in developing countries or in rural areas of the United States, restricting oral contraceptives to doctor's prescription makes this method unavailable for a large proportion of the population. The necessity of a screening pelvic examination has also been questioned.

Of 970 potential oral contraception candidates seen by physicians in London, 44 patients were refused because of their medical history, six patients were refused because of hypertension noted on screening blood pressure, one patient was refused subsequent to a breast examination, and no patients were refused because of the pelvic examination.[59]

The implications for the use of para-medical personnel are evident. Out of sheer necessity the medical profession may have to abandon its traditional roles to accommodate the needs of the society being served.

Statement by the International Planned Parenthood Federation (IPPF) Central Medical Committee[60]

When oral contraceptives were first introduced, it was reasonable to restrict the use of these unknown and relatively powerful drugs to medical prescription. However, as experience has extended over a decade and a half and grown to tens of millions of users, the IPPF Central Medical Committee is increasingly confident that this method of family planning is highly effective and relatively simple to use, and that the health benefits almost certainly outweigh the risks of use in nearly all cases. It has been found that the complications that do occur are difficult to predict by examination prior to use, but that access to follow-up facilities can be important, especially in enhancing continuation rates.

Continuation rates among oral contraceptive users have sometimes been disappointing, but the wide acceptability of the method enables it, by preventing unplanned pregnancy and induced abortion, and permitting the satisfactory spacing of children, to make a contribution towards reducing maternal mortality and increasing the quality of life for parents and their children.

The limitation of oral contraceptive distribution to doctors' prescription makes the method geographically, economically and sometimes culturally inaccessible to many women. As a consequence, deaths and sickness of women and children, which might otherwise be avoided by the voluntary limitations of fertility, continue.

In many countries the regulations that are supposed to limit oral contraceptives to doctors' prescription are generally ignored. Those who can afford to purchase them from commercial outlets do so without medical supervision. However, national and international agencies abide by regulations, only distributing free or subsidized pills through doctors. As a result there is discrimination against many of those most urgently in need of protection against unplanned pregnancy.

The committee recognizes that death due to thrombo-embolic disease is a rare but demonstrable complication of the use of oral contraceptives, and that certain endocrine and metabolic changes take place in some users. Nevertheless the Committee feels that routine examination contributes little to reducing the risks because it feels it is rarely possible to identify susceptible women. The Committee points out that some unknowns remain concerning potential beneficial or harmful long-term side effects, but these are most likely to be elucidated by case control studies which can be carried out independently of the method of distribution. The Committee believes that whoever normally meets the health needs of the community, whether doctor, nurse, traditional midwife, pharmacist, or storekeeper, can be an appropriate person to distribute oral contraceptives. The Committee concludes that responsible, simple methods of non-medical distribution of oral contraceptives can and should be devised and recommends member Associations to:

1. Pioneer innovative schemes for distribution of oral contraceptives (together with all other contraceptives).
2. Educate governments and the medical profession of the health benefits to women and children of non-medical methods of distributing oral contraceptives.
3. Plan programmes of information and education describing the use of oral contraceptives, relative contraindications, and possible side effects.
4. Reorient clinical facilities so that the public has access to trained personnel in cases where the woman is uncertain about the use of oral contraceptives, has a complicating medical condition, or requires reassurance.

Prescriptive Techniques

In order to give first cycle protection, an adjunct may be advisable during the first month of use until the patient familiarizes herself with the routine.

Giving only a six-month's supply at one time may insure follow-up. Unfortunately, there is a disadvantage to this practice also; for certain elements of society this may cause an increase in the discontinuation rate.

Each type of pill has its own special packaging; demonstrate and explain to the patient the one she is using. In the Republic of China hormonal contraception is impregnated on rice paper and supplied as a calendar to facilitate use.

Inform the patient there may be a variant regimen to extend the cycle length to avoid menstruation at inconvenient times.

The progestogen dominant versus estrogen dominant concept is rarely of value to the patient.

Precautions. The following are things the health care provider should take into consideration before prescribing oral contraception and should observe while the patient is on the pill:

1. Pretreatment exam—breasts, bimanual and Pap smear
2. Alteration in menstrual patterns, oligo-ovulation
3. Susceptible blood pressure
4. Alteration of preexisting fibroids
5. Prior psychic depression
6. Altered laboratory studies
7. Cardiac disease
8. Varicose veins
9. Epilepsy, migraine headaches
10. Interraction with other medications, particularly corticoids and certain antiepileptic and antitubercular drugs
11. Asthma
12. Renal dysfunction
13. Women over age 40

Absolute Contraindications. Never prescribe oral contraception to persons who have the following history:

1. Prior thrombophlebitis
2. Prior thromboembolism
3. Marked impaired liver function
4. Known or suspected breast cancer
5. Any estrogen dependent cancer
6. Undiagnosed genital bleeding
7. Congenital hyperlipidemia
8. Breastfeeding at the present time
9. Cerebral vascular disease
10. Coronary occlusion
11. Pregnancy

Reasons for Removing Patient from the Pill. With the occurrence of the following signs, the patient should discontinue use of the pill immediately:

1. Jaundice or pruritus
2. Significantly elevated blood pressure
3. Persistent headaches or migraines
4. Suspicion of breast cancer or estrogen-dependent cancer
5. Persistent or recurrent abnormal genital bleeding
6. Thrombophlebitis, pulmonary embolism, cerebrovascular disorder or retinal thrombosis
7. Visual loss, proptosis, diplopia with papilledema or retinal vascular lesion
8. Amenorrhea for two consecutive cycles unless pregnancy can be ruled out

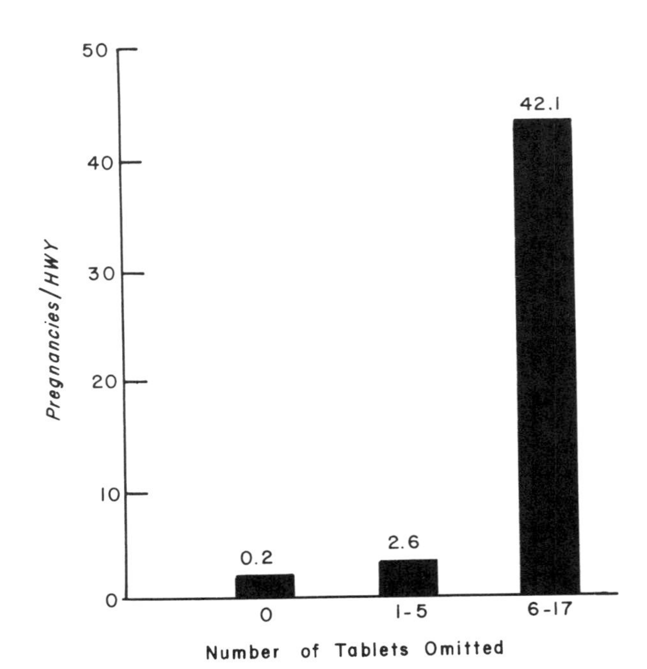

FIGURE 7-35. Pregnancy rate following tablet omission (norethynodrel/mestranol) Enovid 5 mg. with number of tablets omitted. (Based on data from Pincus, G.: *Suppression of ovulation with reference to oral contraceptives*. In Gardiner-Hill, H. (ed.): *Modern Trends in Endocrinology*, Vol. 4, Butterworth Publishing, Inc. Reading MA, 1972, p. 231.)

General Problems and Answers

Problem: Patient calls and wants to know what she should do if she misses a pill.

Answer: Take two pills the next day. If she missed two consecutive pills, take two pills daily for two days and use a condom or foam. Figure 7-35 indicates pregnancy rate following tablet omission.

Problem: Should a lactating woman use the pill?

Answer: Probably not. Minute quantities of steroids and their degradation products are secreted in the breast milk. With much larger doses than in the currently available oral contraceptives there is a resultant suppression in the quantity of milk and a decrease in the percentage of fat and protein. The long term effects on the neonate are unknown. The potential hazards have been signaled by the FDA as a contraindication for the use of the pill in the nursing mother. Biologic evidence is not available.

Problem: Patient would like to know if she is "safe" during the first cycle.

Answer: The patient will be protected after the first week of tablet taking. However, it is frequently advisable to use a concomitant form of contraception during the first cycle until the patient familiarizes herself with the routine. It is recommended, although not essential, that the patient be seen again within two months to insure proper utilization and detection of untoward effects.

Problem: The patient does not want to menstruate during her honeymoon.

Answer: Additional pills may be taken (from an extra pack) to prevent the withdrawal flow. The preceding cycle may also be modified (shortened).

Problem: What pill is suitable for what patient?

Answer: This is difficult to answer. It is difficult to evaluate the hormonal effects of the various pills and therefore the clinical differences between the preparations have not been accurately assessed. In general, mestranol is a weaker estrogen than ethinylestradiol. Some synthetic progestins, especially norethynodrel, have shown, in animal studies, a significant estrogenic effect. Other preparations such as ethynodiol diacetate or higher doses of norethindrone tend to have more anti-estrogenic properties. It is doubtful that claims for the use of such progestogen-estrogen combinations are better suited or tolerated by women who may be estrogenic in character. Unfortunately these data are mostly obtained from animal experiments. The species differences in metabolism of given compounds are great and can be seen in Table 7-17.

Table 7-17. Species differences in metabolism of a given compound.

Species	*Excretion*		*Plasma*
	Urine (%)	*Feces (%)*	*Half-Life (Hours)*
Man	94	1–2	14
Rat	90	2	4–6
Guinea pig	90	5	9
Dog	29	50	23–25
Rhesus monkey	90	2	2–3
Capuchin monkey	45	2	20
Stump-tail monkey	40	60	1
Mini-pig	86	1–2	4–7

Adapted from Christie, G. A.: *Rate limiting factors in the development of new contraceptive methods.* In Potts, M., and Wood, C. (eds.): *New Concepts in Contraception,* University Park Press, Baltimore, 1972.

Clinical experience supports that empiricism still must be relied on in the selection and administration of the particular pill. The only exception is the indication for use of the combined pill over the sequential in patients with heavy menstrual flow.

Summary

Effectiveness: exceedingly high.

Advantages: usage not associated with coital act.

Disadvantages: requires a high degree of motivation or supervision and a certain degree of intelligence.

Risk of failure: low.

Alleged medical risk: low.

Minipills[61,62]

The minipill implies daily hormonal contraception with microdose quantities of 19-nortestosterone or progesterone derivatives. This should not be confused with the low dosage (microdosage) combination oral contraceptives. Terminology between these two types is often confused.

The suggested mechanisms of action in minipills include

1. Reduction of sperm migration through alterations in the physical, biological and biochemical properties of the cervical mucus with possible changes in sperm capacitation.
2. Modification of the endometrial histology and function related to gametes and nidation.
3. Suppression of midcycle LH and FSH peaks in some women (variable effect with suppression of ovulation in 25 percent of patients).
4. Interference with the biosynthesis or metabolism of progesterone by the corpus luteum (luteolysis). Corpora lutea are formed, however, in the majority of patients on the minipill.
5. Possible changes in oviductal secretions and motility.

While unsubstantiated, with elimination of the estrogen component of the combined oral contraceptive there possibly may be fewer hazards related to changes more frequently reported with the combined formulation. These changes include alteration of glucose tolerance, cholesterol or triglycerides; blood pressure elevation; blood coagulation factors changes; and liver stasis.

The major disadvantage of the minipill is the variation in the cycle length with irregular bleeding and changes in the amount of flow and the significant increase in the pregnancy rate when compared with the combined oral contraceptives.

Summary

Effectiveness; high, albeit less than the combination pill.

Advantages: elimination of potential estrogen side effects.

Disadvantages: menstrual spotting; requires high degree of motivation.

Risk of failure: slightly higher than with combined formulations.

Alleged medical risk: possibly lower than with combined pill.

Depo-Provera[63,64,65,66]

Intramuscular injection of a long acting water insoluble progesterone derivative, medroxyprogesterone acetate, in doses of 150 to 300 mgm. at intervals of three to six months, has proven to be an exceedingly effective method of contraception. Higher doses can provide even longer intervals of pregnancy prevention. There are approximately one million users worldwide.

Mechanisms of action suggested are

1. Interference with gonadotropin release, probably at the hypothalmic level with suppression of ovulation.
2. Thinning of the endometrium with a marked suppression of glandular proliferation.
3. Thickening of cervical mucus, decrease in viscosity, absence of both ferning and spinnbarkheit.

Side effects of Depo-Provera include the following:

1. Irregular inappropriate bleeding (Fig. 7-36).

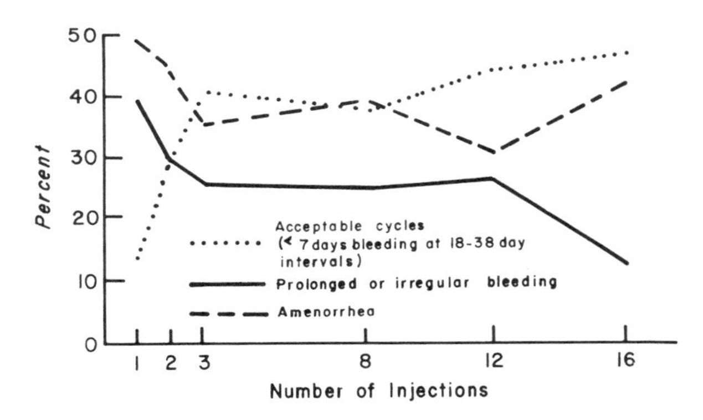

FIGURE 7-36. Percentages of women experiencing various bleeding patterns while receiving three-month injections of Depo-Provera (150 mg.). (From Rinehart, W., and Winter, J.: *Injectable progestogens–officials debate but use increases.* Population Reports, Series K, No. 1, 1975, with permission.)

2. Amenorrhea. This is of unpredictable onset. It is present in 23 percent of patients after one month of therapy, 69 percent at two years, and 80 percent of the patients by three years.

3. After cessation of treatment 50 percent of the patients will have regular menses initially while 75 percent will have regular menses after one year. A significant number of patients will have prolonged menstrual irregularity. It is the severity and difficulty in management of the menstrual irregularities when they do occur that are discouraging in the use of this approach.

4. Prolonged infertility and possibility of permanent infertility. Of those attempting to conceive following cessation of treatment, 66 percent will do so by nine months and 75 percent by one year. Of those patients who conceived after cessation of therapy, the average time to conception from the last shot was 11 months with a range of 4 to 31 months (Fig. 7-37).

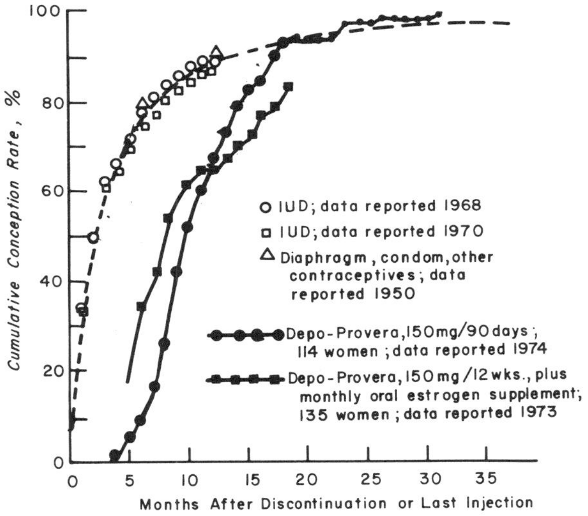

FIGURE 7-37. Cumulative conception rates for women who discontinued Depo-Provera to become pregnant, compared with cumulative conception rates for former users of IUD's and of traditional contraceptives. (Adapted from Rinehart, W., and Winter, J.: *Injectable progestogens–officials debate but use increases*. Population Reports, Series K, No. 1, 1975, with permission.)

5. Slight weight gain noted in majority of patients. There are no reported changes in glucose tolerance, serum triglycerides or free fatty acids, liver function studies or blood pressure.

There is a small change in blood coagulation factors and blood estrogen levels (mean estradiol levels comparable to those in the early proliferative phase).

The long term effects on target organs have not been defined. Comparisons with the oral contraceptive may be misleading, if only because of the restricted differences in population sampling.

Problem: Should the method be used in lactating women?

Answer: No. Depo-Provera has no adverse effect and a possible beneficial effect on lactation; however, the effects on the neonate are not known and it is not approved by the FDA for use during lactation.[67]

This drug has a public health advantage in that (1) it has a carry-over effect, (2) it requires infrequent administration, (3) it need not be administered by a physician and (4) it is particularly helpful in those individuals incapable of using other contraceptive methods.

Summary

Effectiveness: exceedingly high.

Advantages: minimal reliance on patient; devoid of coital act; public health aspects; simple to administer.

Disadvantages: menstrual problems, infertility.

Risk of failure: very low.

Alleged medical risk: low.

Pill-A-Month

Quinestrol is a synthetic estrogen, which is stored in adipose tissue and slowly released into the general circulation. It is often combined with the synthetic progestin quingestanol (Fig. 7-38). The erratic and unpredictable rate of release can lead to difficulty in establishing a cyclic pattern and thus can lead to contraceptive failure.

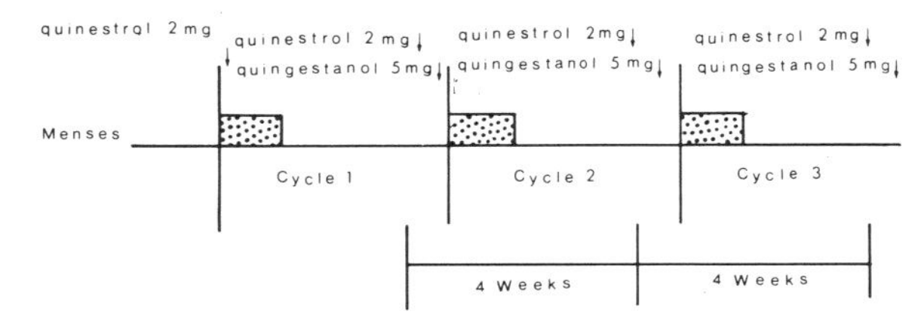

FIGURE 7-38. Schedule of administration of one pill-a-month. (Adapted from Guiloff, E., et al.: *Clinical study of a once-a-month oral contraceptive quinestrol-quingestanol.* Fertil. Steril. 21:110, 1970.)

Summary

Effectiveness: lower than combined.

Advantages: obvious, one pill per month.

Disadvantages: same as for other oral contraceptives plus slightly more frequent undesired vaginal bleeding episodes.

Risk of failure: greater than that of combined oral contraceptives.

Alleged medical risk: same as for combined oral contraceptives.

Polymer Implants

Intravaginal, intrauterine, subcutaneous or subdermal implants of an inert silastic device containing a progestogen derivative have been utilized in preventing pregnancy. Depending on the hormonal agent used, the device has to be replaced at varying intervals, but can be effective for up to five years. The mechanism of action relies upon the slow release of the hormone, as can be seen from the increase of the basal body temperature and suppression of the midcycle LH surge.

These devices in general are easy to implant and remove. There is no conclusive evidence that any of these devices have any significant advantages over available IUD's. They may be useful, however, for replacement therapy in hormonal deficiency states. Early trials using androgens for contraceptive action are underway in males.

Summary

Effectiveness: moderately reliable.

Advantages: long acting slow release of progestogens.

Disadvantages: subcutaneous implants and erratic cycles. At present time this must be considered an experimental approach.

Risk of failure: studies underway suggest promise of low failure.

Alleged medical risk: low.

Morning-After Pill[68,69,70]

The morning-after pill is also known as postcoital contraception, pregnancy interception and postovulatory contraception.

The morning-after pill implies the utilization of high doses of estrogens given over several days. The effectiveness is directly related to the amount given. Some of the more common schedules include:

diethylstilbestrol 25-50 mgm. b.i.d. x 5 days
conjugated estrogens 5-10 mgm. t.i.d. x 5 days
ethinylestradiol 1-5 mgm. q.i.d. x 5 days

The possible mechanism of action includes (1) prevention of nidation by an alteration in endometrial histology and chemistry, (2) acceleration of tubal transport of the ovum and (3) corpus luteum suppression.

The morning-after pill can be given within 72 hours following an unprotected exposure. In those patients in whom this was the only exposure during that cycle the pregnancy rate is very low (less than 1 percent). The morning-after pill will not interrupt an implanted pregnancy.

Side effects are noted by more than 50 percent of patients. None are serious. They include nausea and vomiting, headache, menstrual disturbance and breast tenderness.

Even though the exposure is prior to the time of organogenesis of the female reproductive tract, which is 5 to 6 weeks (see Fig. 7-33), and despite the fact that teratogenic effects have not been noted in either the monkeys or the human, the presence of an unexpected preexisting pregnancy must be considered. Possible diethylstilbestrol-induced lesions in the vagina

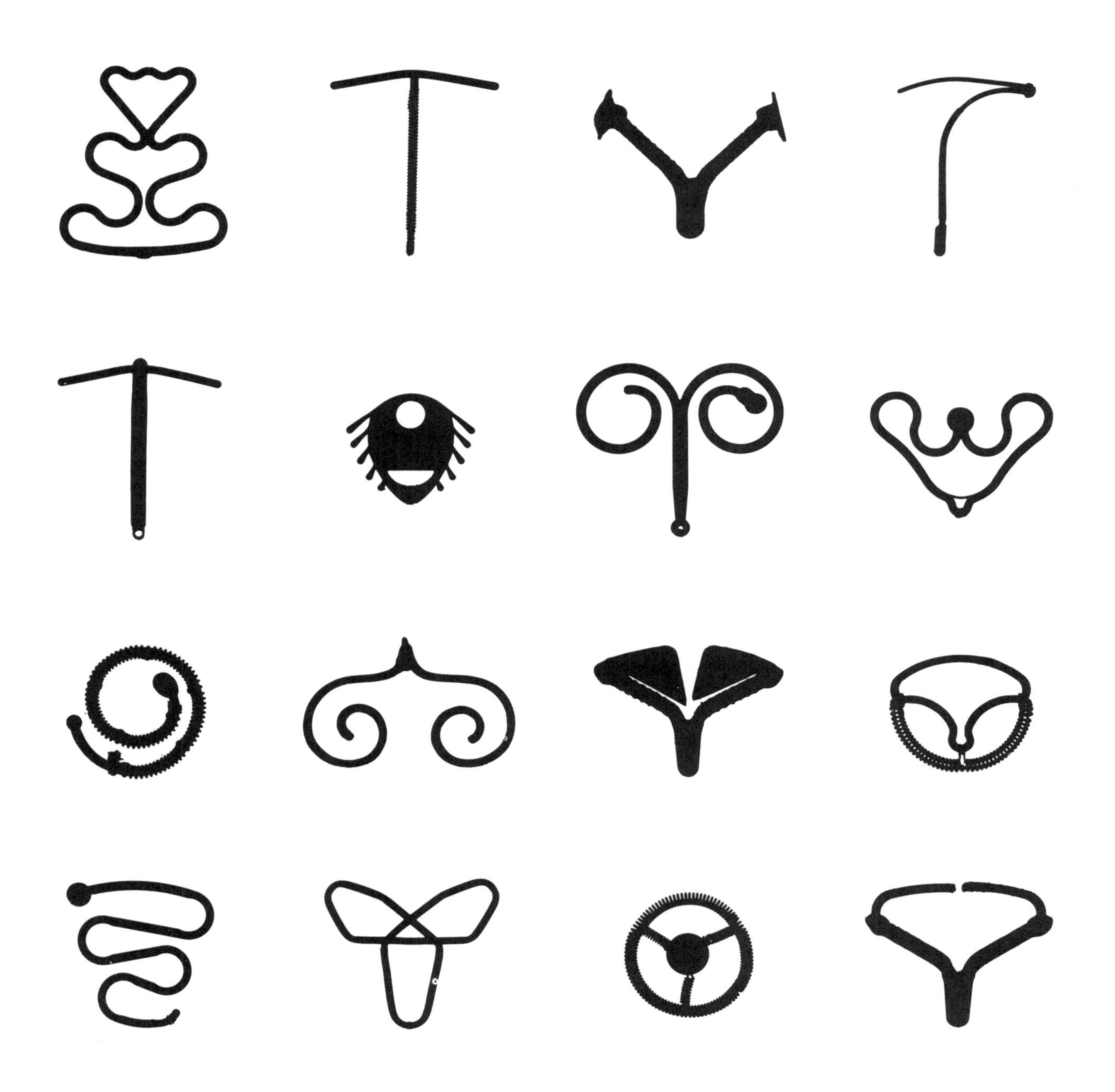

FIGURE 7-39. A variety of intrauterine devices. (Design by Rudolph de Harak, with permission and by courtesy of Family Planning Perspectives.)

of female fetuses must be avoided and abortion counselling should be offered for failures.

Summary

Effectiveness: moderately reliable (FDA accepts its effectiveness).

Advantages: the only approach for unplanned exposure; should be used only as an emergency method, e.g., rape.

Disadvantages: side effects (nausea and bleeding).

Risk of failure: low—should not be used as routine contraception or used repeatedly.

Alleged medical risk: (1) same as oral contraceptives; (2) diethylstilbestrol potentially teratogenic; (3) effects on fetus poorly understood; and (4) if pregnancy occurs in spite of use, voluntary pregnancy termination should be offered.

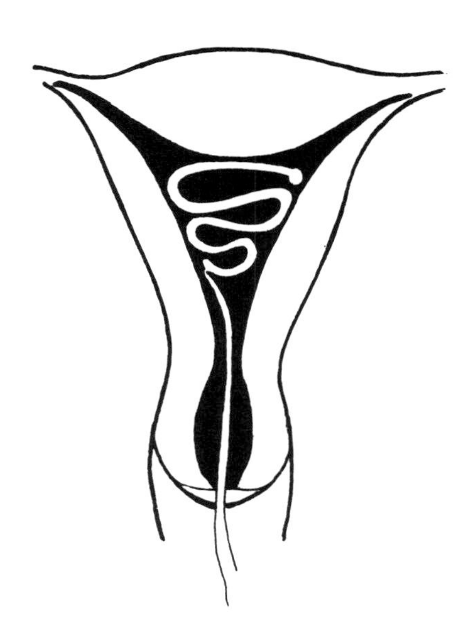

FIGURE 7-40. An intrauterine device, a Lippes loop, in utero.

Intrauterine Device (IUD)

Although the concept of the intrauterine device dates back to ancient times when stones were inserted into the uteri of camels to prevent pregnancy during a prolonged desert crossing, it was not adapted for appreciable human use until the Graffenberg ring was introduced in 1928. This device consisted of a silver wire and coincidentally contained a significant amount of copper. However, use of this device still was limited and generally condemned until the mid-1950's when improved devices were introduced and the inherent advantages of the contraceptive modality were recognized. Today there are over 15 million women using the IUD worldwide (Figs. 7-39 and 7-40).

Despite the numerous varieties of IUD's no one type can be universally recommended for all women.

The mode of action is unknown. The endometrium is infiltrated by leukocytes, producing an endometrial exudate. This may function to be spermicidal, prevent fertilization, be blastotoxic or inhibit implantation. There is no effect on ovulation, ovarian steroids or gonadotropins or tubal transport. The latter is questionable because tubal motility studies are lacking.

Continuation Rates

IUD continuation rates are generally reported to be better than the pill. However there are many factors which could affect patient motivation and which could be cited to explain the variance in continuation rates at different times and in different countries. Continuation rates do not usually take into account the age and parity of the users. IUD users are generally older than oral contraceptive users (see Fig. 7-7). The IUD method is more frequently chosen for limitation while oral contraceptives are chosen for spacing.

Removal and expulsion rates have been reported to decline with parity within each age group. The increased acceptance of sterilization among older couples (over 30) may have an effect in changing these rates (Fig. 7-41).

Pregnancy, expulsion and removal rates for bleeding and/or pain are inversely related to the size of the IUD. (Fig. 7-42)

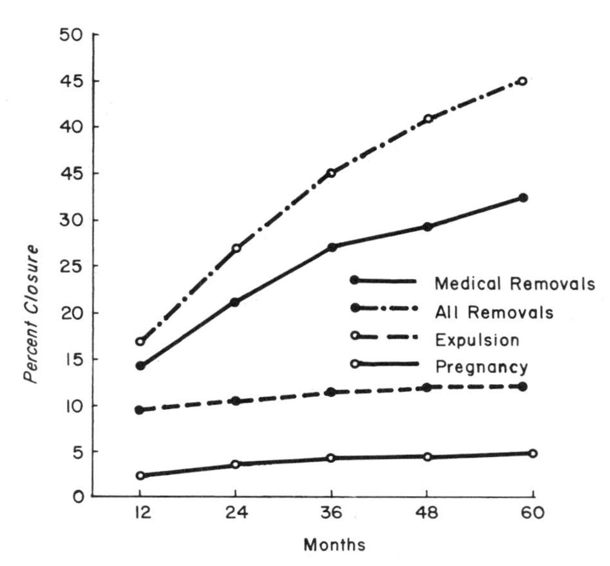

FIGURE 7-41. Reasons for discontinuation of IUD. (Data from Tietze, C.: *Evaluation of intrauterine devices: ninth progress report of the Cooperative Statistical Program.* Studies Fam. Plann. 1:1, 1970.)

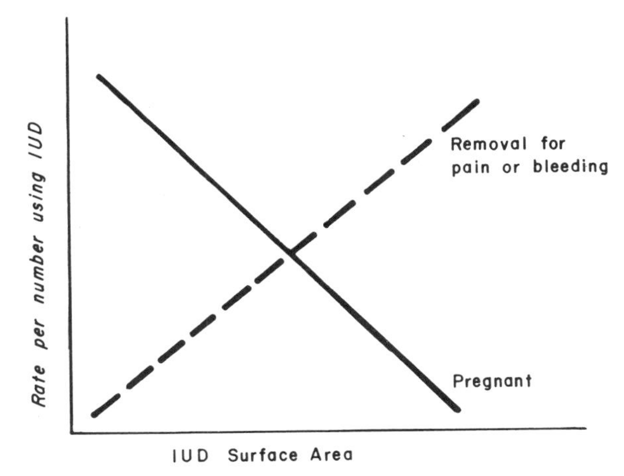

FIGURE 7-42. Relationship of IUD removal and failures with surface area of IUD.

Side Effects

Because of inconsistent reporting and difficulty in comparing results of different studies, IUD complication rates are difficult to define. Nevertheless the following can be determined.

1. Uterine perforation—most commonly occurs with insertion. Erosion through the myometrium is less likely although it can occur. Imprint of the device in the endometrial surface can be readily seen. The incidence is 1 to 7/10,000 and related to the experience of the physician in inserting the IUD. Incidence can be reduced by (1) careful pelvic examination prior to insertion to assess position of uterus and (2) uterine sounding prior to insertion to assess direction of the cavity.

Problem: How can an IUD be removed from the peritoneal cavity following perforation?
Answer: The IUD can frequently be easily removed through the laparoscope. This is not the case with the copper IUD or the Dakon Shield since omentum with increased vascularity tends to envelop the IUD. Removal of these devices through the laparoscope may be more hazardous than for example the Lippes loop.

Problem: The patient cannot feel the strings on her IUD. What should she do?
Answer: The IUD strings should be checked prior to each coitus. If they cannot be located by the patient, she should be instructed to use another contraceptive method, i.e., foam, condom or diaphragm, until she can be examined. In most instances the IUD strings will be seen by the examiner and the patient can be reassured.

In those cases where it is not seen, the uterine cavity can be gently probed with a hook or an endometrial biopsy curette. In those few instances where the IUD cannot be found there are several additional techniques for IUD localization. These include anteroposterior and lateral x-ray study following insertion of a second IUD of another type; x-ray following insertion of a radiopaque dye; x-ray after insertion of a uterine sound or probe; ultrasonography; and hysteroscopy.

2. Expulsion of IUD. This is inversely related to the size of the IUD and age and parity of the patient (see Fig. 7-42); approximately 50 percent of these patients will retain their IUD

with reinsertion. Rates decline with successive months of use. As many as 20 percent of expulsions may go unnoticed.[71]

3. Pregnancy—rate 2 to 3 percent. Generally pregnancy is associated with expulsion or displacement and rates decline with successive years of use. The frequency of ectopic pregnancy, especially ovarian, is higher for women who become pregnant with the IUD. That relates to the relative failure of the IUD to prevent extrauterine pregnancies.

4. Bleeding. This is the medical reason most often accounting for IUD removal. Menses tend to be heavier and longer and occur two days earlier than a comparable time in control cycles as a result of the local effect on the endometrium. Furthermore, intramenstrual spotting is not uncommon. In normal cycles 35 cc. of blood are lost; with the standard IUD 70 to 80 cc. are lost; while with the copper T 50 to 60 cc. are lost. Following discontinuation of the IUD, the menstrual flow may not return to pre-IUD patterns for up to six months. Anemia could become a problem, especially for women in developing countries.

5. Dysmenorrhea—may be aggravated by an IUD. Patients with a history of troublesome dysmenorrhea generally are bad candidates for this method.

6. Pelvic pain—seen in 2 to 3 percent of patients with the IUD.

7. Infection. Although adequate data is not available (most studies have been done with patients who may be prone to high rates of pelvic infections), the following has been determined: (a) there is an occasional reactivation of chronic pelvic inflammatory disease with insertion; (b) a retrospective study has suggested a ninefold increased risk of acute pelvic infection among IUD wearers as opposed to nonwearers[72]; (c) carefully obtained transfundal cultures have revealed a sterile endometrium generally within 24 hours after insertion; and (d) incidence of postinsertion infection is low (1 to 2 percent). Infection may be reduced by (a) strict adherence to sterile technique on insertion; (b) avoidance of insertion in women with recent histories of pelvic inflammatory disease; and (c) preinsertion gonococcal culture.

General Problems and Answers

Problem: Should the IUD be removed if pelvic inflammatory disease occurs?

Answer: If the patient responds to antibiotic therapy the IUD may be left in situ.[73]

Problem: Can the male feel the strings during coitus?

Answer: No. However, if the strings are too long they can become entwined with pubic hair fragments or a tampon, thus causing an irritation to both partners. If the strings are too long, they can be excised without removal of the IUD.

Problem: What should be done if a patient becomes pregnant with an IUD?

Answer: Remove the IUD. The abortion rate with the IUD in situ is approximately 50 percent; when removed it is 30 percent.[74] There are numerous reports in the literature of severe puerperal infection and maternal deaths occurring in the second trimester in patients with the IUD left in situ.[75] If the IUD cannot be removed easily without intrauterine intervention, pregnancy termination should be discussed with the patient and the IUD should be removed in the operating room. Teratogenic effects of the IUD have not been documented.

Problem: Can the IUD be inserted during the early postpartum or postabortal period?

Answer: Yes, unless there is evidence of concomitant infection. The advantages to early postpartum and postabortal insertion are (1) substantial demographic benefit; (2) increase in contraceptive utilization; and (3) no increase in the rate of infection, pain, abnormal bleeding or perforation. Disadvantages are (1) increase in the expulsion rate for postabortal insertions and (2) possible increase in the pregnancy rate.[76]

Problem: Are there contraindications to insertion of the IUD?

Answer: Yes—pregnancy, pelvic inflammatory disease, undiagnosed vaginal bleeding, severe anemia, congenital uterine anomalies, uterine cavity distorted by fibroids and severe flexion with fixation of fundus.

Problem: Which IUD's are best suited for the nulligravida?

Answer: Saf-T-Coil, Progestosert® or Copper 7. These are the smallest.

Insertion and Retention

Each type of IUD has minor differences in the technique of insertion but all are similar to uterine sounding. Familiarity with the particular IUD to be inserted is essential. Suggestions for improvement in the ease of insertion and retention are the following:

1. Examine the patient before insertion to rule out pregnancy or pelvic inflammatory disease and to ascertain the position of the uterus.
2. Use tenaculum traction to help straighten the axis of the uterus.
3. Insertions are easiest at the time of the menses. This also avoids the hazards of insertion during an early pregnancy.
4. Insertion may be facilitated with a paracervical block or Valium premedication.
5. Vasovagal reactions are not uncommon and can be avoided with preinsertion atropine.
6. Administration of a strong analgesic such as morphine or pantapon given immediately postinsertion will increase the number of retentions but need be resorted to infrequently. Facilities should be available within the clinic or arrangements should be made with friends for careful observation of the patient.
7. Do not cut the strings too short since they tend to retract somewhat into the uterus following insertion and the patient may have difficulty finding them.

Copper IUD

Recent advances in IUD technology have produced the copper IUD's. These consist of a plastic base (generally appearing in the United States in the shape of a T or a 7) upon which is wound a coil of copper wire with a surface area of some 200 mm^2. Copper is released from this device at a rate of 50 mcg. per day, far below the toxic daily doses of 1 mgm. per day. Serum and tissue copper levels in women wearing this device is essentially unchanged.

Animal experiments have shown the device to be effective locally only. Other possible mechanisms of action are the following[77]:

1. Inhibition of cleavage and nidation of blastocysts—questionable. Fertilized eggs from rats or rabbits with copper wire in place when transferred to a recipient develop normally.
2. Spermicidal—of questionable importance. Pregnancies occur on rare occasion with device in place.
3. Increase in endometrial leukocytes—more so than with conventional IUD's. Anti-inflammatory agents such as indomethacin may reduce the effectiveness of the copper IUD in experimental animals.
4. Change in endometrial histochemistry.
5. Cytotoxicity to blastocysts shown in mice in vitro but contraindicated in transfer experiments cited in number 1 above.
6. Increase in uterine prostaglandins.

Advantages to the copper IUD are (1) better acceptance among nulliparas because of decrease in pain on insertion, decrease in expulsion rate and increase of continuation rate; (2) decreased incidence of abnormal bleeding and heavy menses; (3) no increase in pregnancy rate[78]; and (4) may protect against gonorrheal salpingitis.

Problem: A patient would like to know how often her IUD should be changed.

Answer: There is no reason to change a conventional IUD unless the patient has symptoms of staining or bleeding after several years of use, potentially rendering the IUD as a reactive foreign body. However, since copper is lost by ionization and chelation, the copper IUD'S should be replaced every two years. Newer models with increasing amounts of copper may be retained for longer intervals. However, this may be associated with higher rates of expulsion. Calcium deposits have been noted occasionally to develop on conventional IUD's after several years in utero. This may cause some local effects.

Intrauterine Progesterone Contraceptive System

The Progestasert®* is a T-shaped IUD constructed of ethylene/vinyl/acetate (EVA) copolymer and containing a reservoir of 38 mg. of progesterone (USP) which is released at an average rate of 65 μgm./day into the uterine cavity. The Progestasert has a local effect on the endometrium with no effect on ovulation, blood steroid levels or blood chemistry levels related to liver, kidney and thyroid function. The mechanism of action is unclear. The efficacy of the T-shaped device without progesterone is notably reduced. Full clinical experience is somewhat limited in that the device has only recently been introduced commercially. Disadvantages include the need for yearly replacement and the increased cost. A comparison of the efficacy of this method over the conventional IUD's is still limited.

Summary

Effectiveness: very high.

Advantages: no coital interference; requires little motivation or intelligence.

Disadvantages: dysmenorrhea, bleeding, pain, expulsion, perforation, and pregnancy related complications.

Risk of failure: low.

Alleged medical risk: low, but with an occasional flare-up of pelvic inflammatory disease and increased severe intrauterine infections during pregnancy with the IUD in place.

Sterilization (Surgical Contraception)

The requests for contraceptive sterilization in this country and through the world have been increasing markedly (Fig. 7-43). In counselling these patients the physician must be aware of the patient's motivation and understanding of the procedure. Does she desire to delay a wanted pregnancy or is she contracepting for the purpose of birth limitation? There is a span of some 15 to 25 years in the average fecund female between the time of her last pregnancy and menopause. Consider the following:

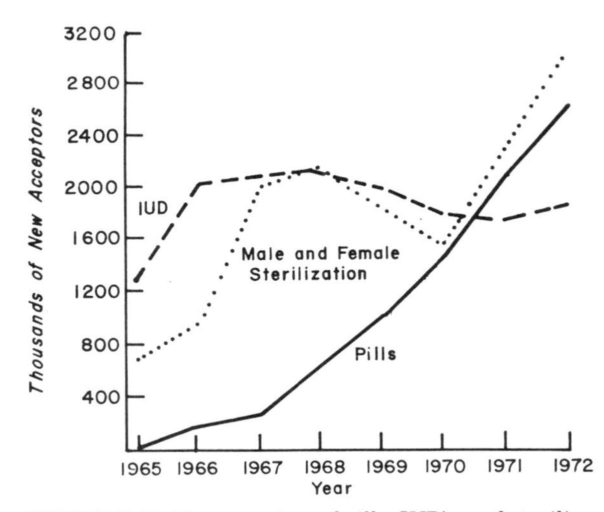

FIGURE 7-43. New acceptors of pills, IUD's, and sterilization in family planning programs in developing countries, 1965 to 1972. (From Piotrow, P., *Oral Contraceptives—50 million users.* Population Report, Series A., No. 1, 1974 with permission.)

1. The span of 15 to 25 years would thus require prolonged contraception with an overall failure rate in American couples of approximately 30 percent.
2. Although the average woman would have to consume over 5000 birth control pills, experience has indicated that most women will stop the pill before this time.
3. A reduction in the infant-childhood mortality rates, particularly in the developed countries, has permitted the patient to achieve her desired family size after fewer pregnancies.

For these reasons and others, sterilization has been chosen by nearly one couple in five in this country with a better than 50 percent increase since 1965. Previously utilized for strict medical and eugenic indications, sterilization is now used as an equally competing form of contraception.

Incidence. The official manual of the American College of Obstetrician and Gynecologists up until 1969 suggested that sterilization be performed in females if they were at least 25

*Registered by the Alza Corporation, Palo Alto, Ca, 1975.

years old with 5 living children, 30 years old with 4 living children or 35 years old with 3 living children. There were approximately 4 million sterilized couples in the United States by 1974 (Fig. 7-44). In that year over 1.3 mil-

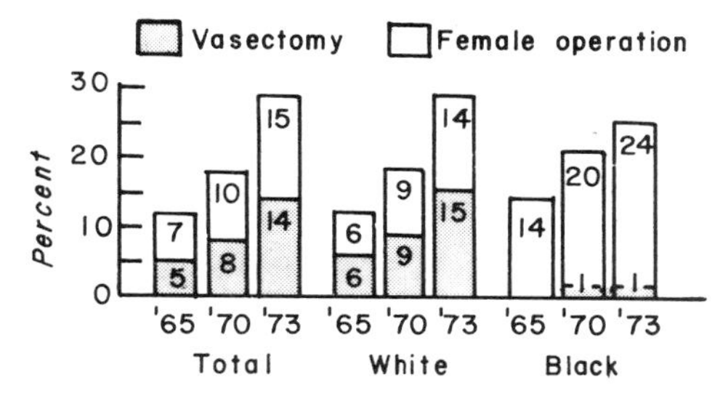

FIGURE 7-44. Percent of U.S. currently married women at risk of unwanted conception reporting that they or their husbands have had a contraceptive sterilization, by type of sterilization and race, 1965, 1970, and 1973. (Adapted from *Digest: One-fifth of U.S. Couples—more than 7 million—rely on contraceptive sterilization; procedures doubled in 4 years.* Fam. Plann. Perspect. 7, No. 3, May/June 1975.)

lion couples selected sterilization, an increase of over 43 percent over the previous year. Almost half those sterilized were women. In 1970 four times as many men as women were sterilized.[79] Nonetheless this is only a small percentage of the total population at risk. Table 7-18 indicates the accumulated sterilizations performed in selected countries through National Family Planning Programs by 1969.

Table 7-18. Accumulated sterilizations performed through National Family Planning Programs in selected countries as of 1969.

Country	*No. of sterilizations in thousands*	*Adjusted no. of sterilizations per 100 married women aged 15–44*
India	5200	5
Korea	118	3
Tunisia	4	0.4
Taiwan	1	0.1

Adapted from Presser, H. B.: *Voluntary sterilization: a world view.* Reports on Population/Family Planning No. 5, July 1970, p. 11.

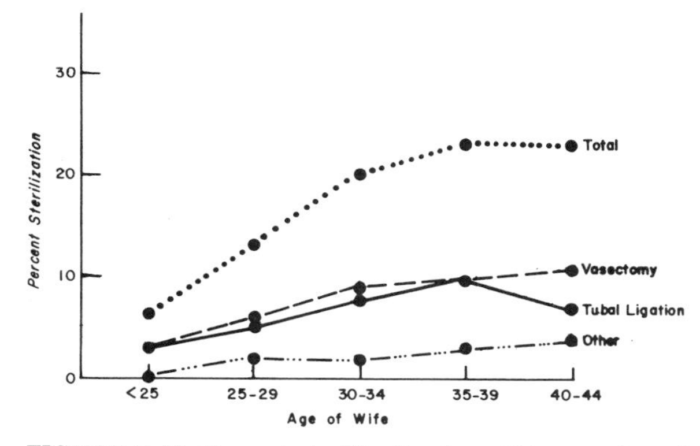

FIGURE 7-45. Percent sterilization by wife's age. (Based on data from Presser, H.B., and Bumpass, L.L.: *The acceptability of contraceptive sterilization among U.S. couples: 1970.* Fam. Plann. Perspect. 4:18, 1972.)

Age (Fig. 7-45). By 1970 sterilization was the first choice of contracepting couples where the wife was 30 to 44 (25 percent for sterilization and 21 percent for the pill). The median age of couples sterilized since 1965 is females, 31.8, and males, 35.1. Forty percent of couples adopted sterilization before the woman reached 30 (Fig. 7-46).

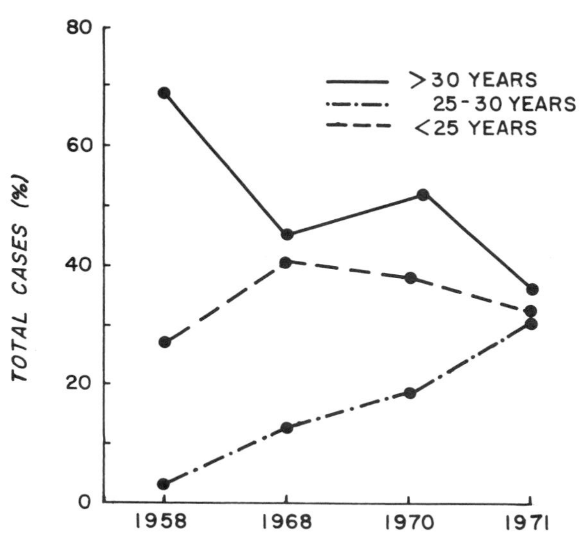

FIGURE 7-46. Change in the age range of women seeking sterilization, 1958 to 1971. (Based on data from Edwards, L. E., and Hakenson, E. Y.: *Changing status of tubal sterilization: an evaluation of fourteen years' experience.* Am. J. Obstet. Gynecol. 115:347, 1973.)

Table 7-19. Reported failure of tubal sterilization for ten specified countries (limited to studies published in 1960 or later).

Country	*Technique for tubal sterilization*	*Failures per 100 cases*
Ceylon	Pfanenstiel incision	0.0
England	Not indicated	0.0
	Pomeroy and Irving	0.0
Finland	Isthmic-tubal resection	0.2
Hong Kong	Pomeroy	0.4
India	Pomeroy and Madelener	0.1
	Pomeroy	0.4
	Not indicated	1.9
	Pomeroy	0.2
Japan	Uchida	0.0
Puerto Rico	Not indicated	1.0
Scotland	Modified Pomeroy	0.0
Switzerland	Leghardt	0.1
United States	Cook	0.4
	Pomeroy	0.0
	Pomeroy	2.0
	Pomeroy and others	0.6
	Fimbriectomy	0.0
All Studies		0.1

Modified from Presser, H. B., and Bumpass, L. L.: *The acceptability of contraceptive sterilization among U.S. couples: 1970.* Fam. Plann. Perspect. 4:18, 1972.

Parity. Of those couples adopting contraceptive sterilization since 1965, the median parity was 2.8 children. The average parity of females accepting sterilization today is nearly half that of ten years ago. Sterilization is still accepted by a larger percentage of high parity couples than low.

Female Sterilization

There are a large variety of techniques for sterilization of the female. For those procedures which do not involve the removal of the uterus, the serious morbidity is very low and the failure rate less than 1 percent (Table 7-19). The overall complication rate, morbidity and failure rate for all techniques of female sterilization are listed in Table 7-20. Several of the more commonly used techniques will be touched upon in this section.

Abdominal Tubal Ligation. Figure 7-47 shows various surgical tubal contraceptive techniques. The advantages to abdominal tubal ligation are

1. can be done easily immediately postpartum under local anesthesia and during cesarean section.
2. good control over the amount of tube removed and the portion removed.
3. very low failure rate.

Table 7-20. Composite data of currently practiced female sterilization techniques.

Procedure*	*Complications (%)* Range	*Complications (%)* Mean†	*Morbidity (%)* Range	*Morbidity (%)* Mean†	*Number of deaths (%)*	*Failures (%)* Range	*Failures (%)* Mean†
Puerperal abdominal TL	0–18.5	1.9	0.6–20.7	8.4	0	0–1.00	0.2
C-section BTL	0–51.1	10.3	5–33.3	19.7	1	0–2.3	0.3
Interval abdominal BTL	0–39.1	5.6	0–7.4	3.6	0	0.5	0.6
TA-abdominal TL	0–41.5	30.8	7–26.0	21.3	0	0	0
Interval laparoscopy BTL	0–8.0	1.3	0–2.9	0.6	2.012	0.1	0.3
Puerperal laparoscopy	0–2.0	1.07	0–3.2	2.6	0	0.1	0.1
TA-laparoscopy TL	0–2.0	0.7	0–4.0	1.5	0	0.3	0.1
Interval vaginal TL	0–26.3	3.7	1.4–20.8	6.9	0	0.3	0.3
TA-vaginal TL	0–41.7	2.4	0.25–33.3	4.4	0	0–0.25	0.1
Chemical occlusion	0–7.0	4.5	0–50.0	2.3	0	3–12.8	5.6
Metal clips	0–2.8	0.5	0	0	0	0–27.11	1.2
Hysteroscopy	0–9.9	3.3	0	0	0	0–5.6	2.3
Interval abdominal hysterectomy	2.3–54.5	12.3	0–18.1	10.8	0	0	0
C-section hysterectomy	0–71.4	15.7	0–50.3	42.4	14.4	0	0
TA-abdominal hysterectomy	0–61.6	29.4	5–43.5	32.4	0	0	0
Puerperal hysterectomy	3–71.4	14.0	21.5–50.8	37.8	0.4	0	0
Interval vaginal hysterectomy	0.9–63.8	23.8	10.5–79	40	0.2	0	0
TA-vaginal hysterectomy	1.5–26.0	8.0	18–25	19	0	0	0

Adapted from Shepard, M. K.: *Female contraceptive sterilization.* Obstet. Gynecol. Survey, 29:739, 1974.

*Key to procedures: TL–tubal ligation
BTL–bilateral tubal ligation
TA–therapeutic abortion

†Means are reported only from those series reporting complications and morbidity.

Disadvantages are

1. For nonpuerperal patient, it requires general anesthesia, an abdominal incision and approximately four days hospitalization.
2. Fimbriectomies are nonreversible.
3. There can be formation of hydrosalpinx and recanalization following fibriectomies in which all tissue proximal to the ampullary isthmic junction is not removed.

Vaginal Tubal Ligation. A Pomeroy tubal ligation or fimbriectomy can be done through a posterior colpotomy incision. The advantages to the procedure are

1. leaves no abdominal scar
2. occasionally can be done under local anesthesia
3. shortened hospitalization

Disadvantages are

1. increased infection rate, especially pelvic abscesses
2. may be difficult to control hemorrhage
3. occasional dyspareunia after the procedure
4. performed with difficulty in presence of pelvic adhesions, enlarged uterus or other pelvic pathology
5. requires adequate exposure and facilitation with special instrumentation

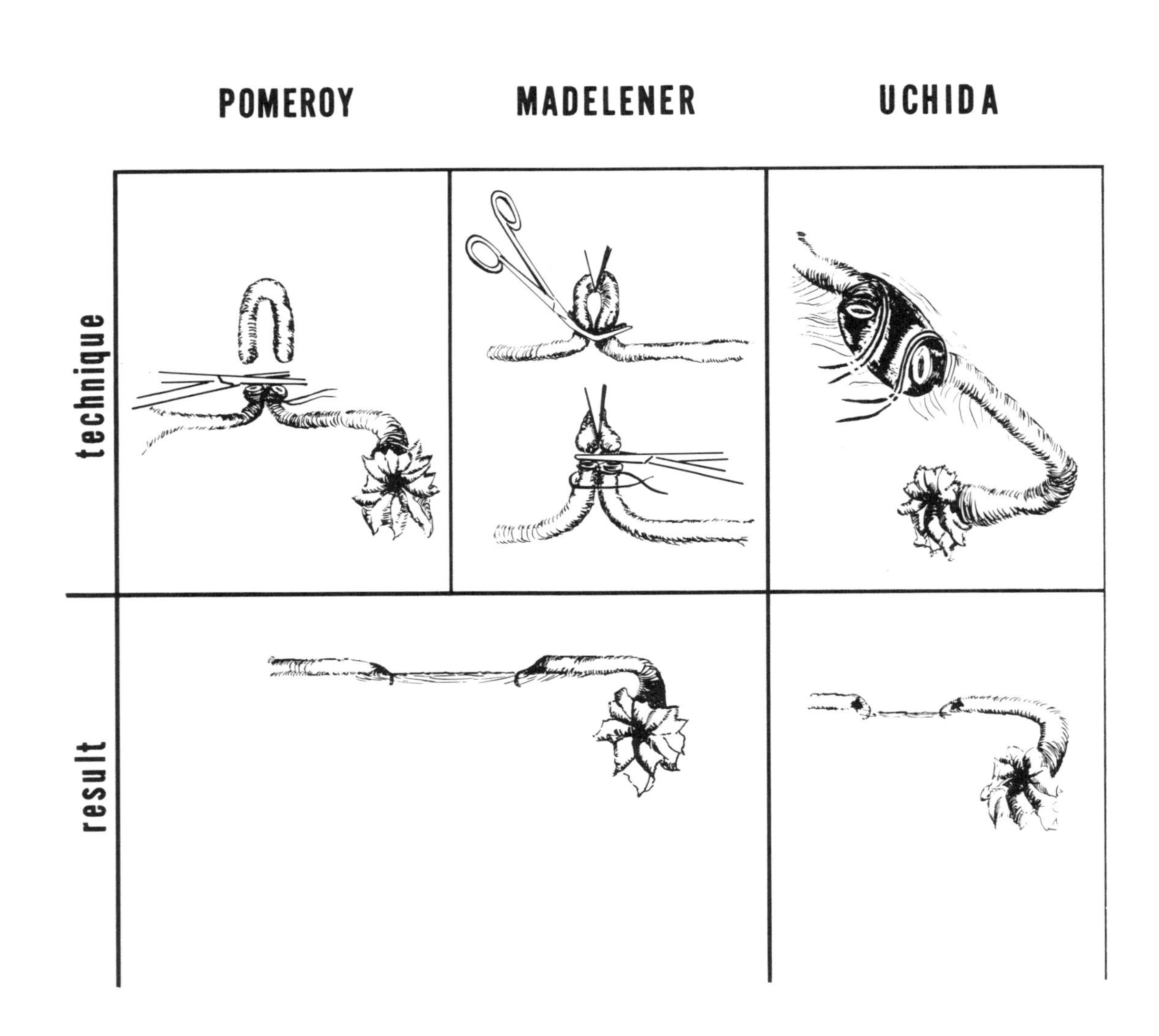

FIGURE 7-47. Surgical tubal ligations and their results.

FIMBRIECTOMY IRVING ALDRIDGE

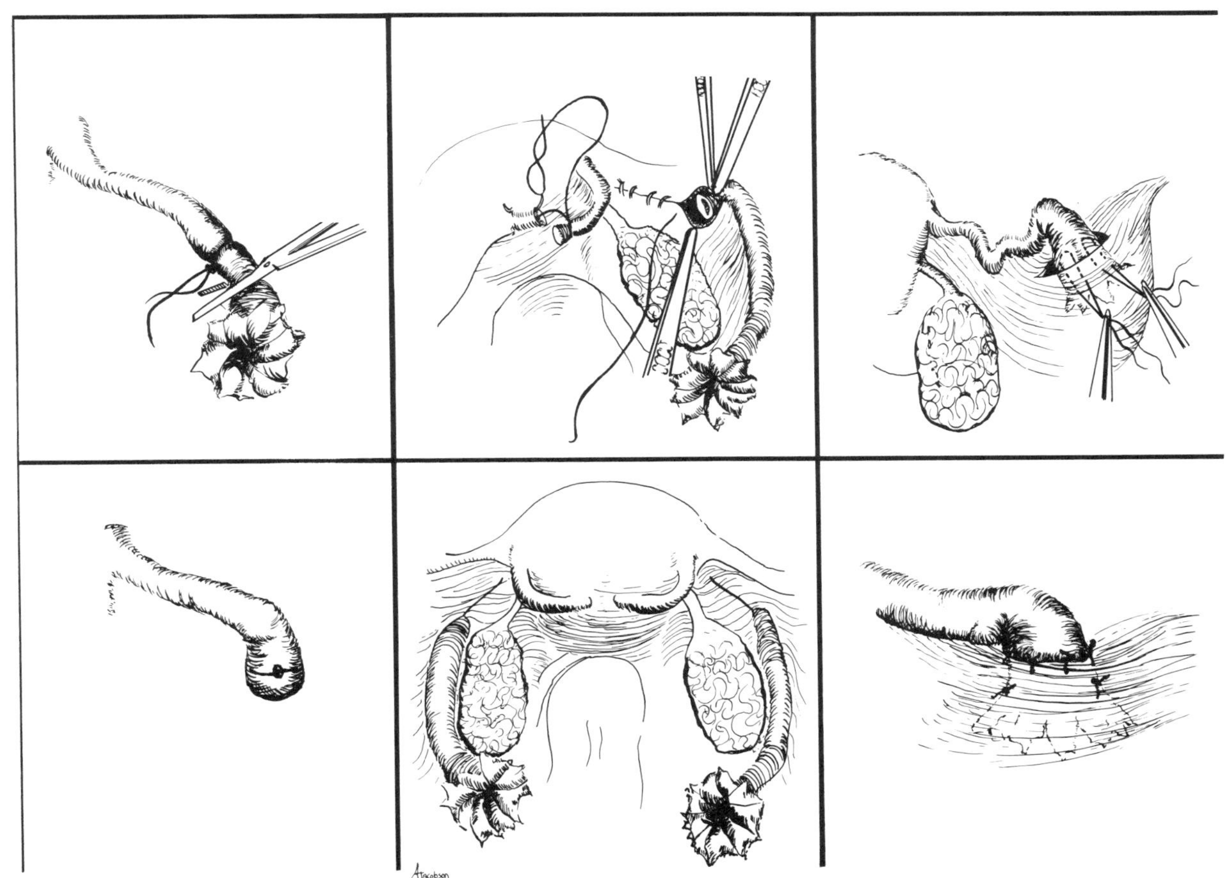

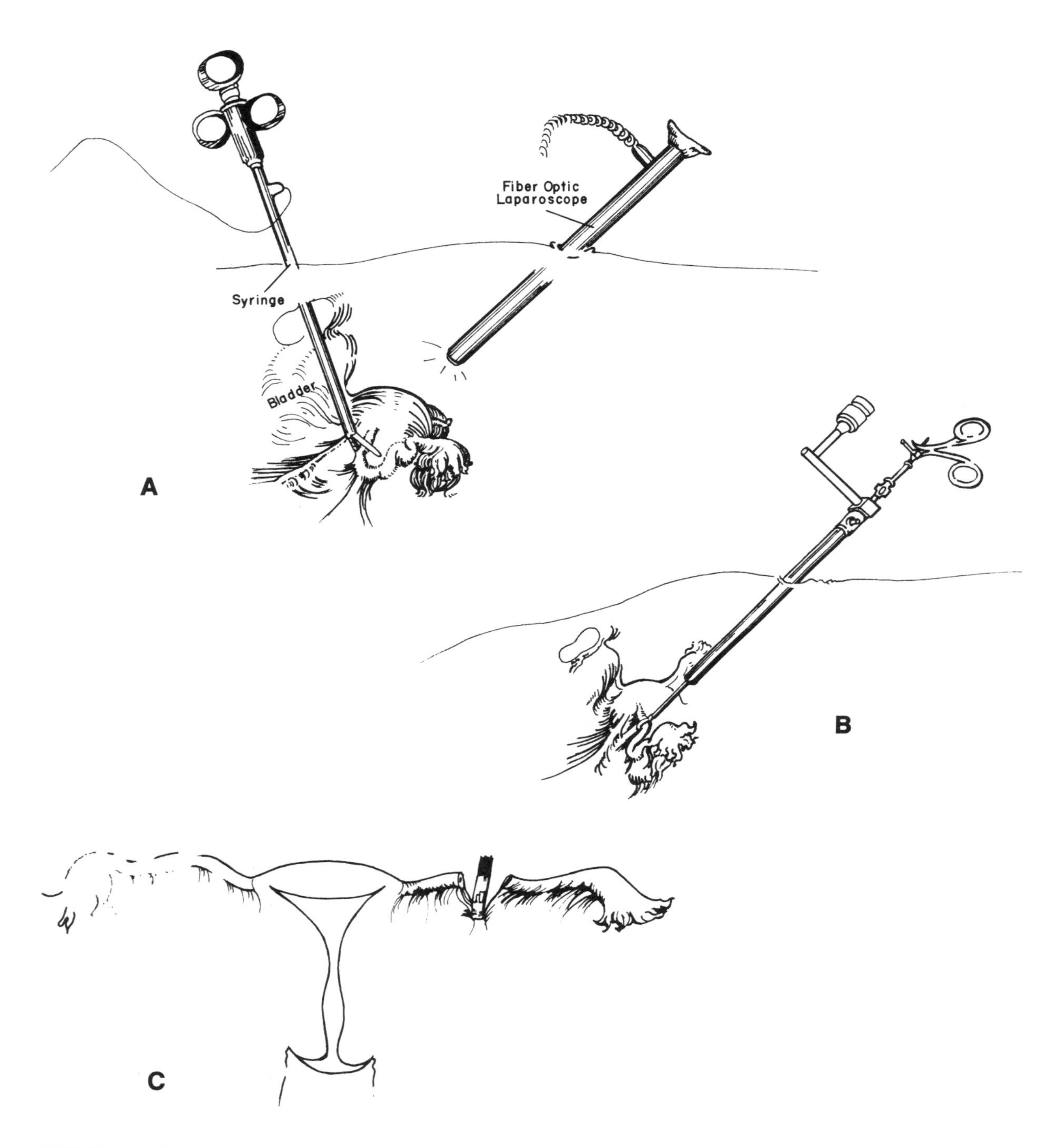

FIGURE 7-48. Laparoscopy sterilization. *A*, Traditional two-incision technique; *B*, One-incision technique; *C*, End result of coagulated tube.

Laparoscopy Tubal Ligation. This is known by the public as "belly-button" or "band-aid" surgery. The tubes are cauterized and/or cut with unipolar or preferably bipolar current under direct visualization with the fiberoptic laparoscope (Fig. 7-48).

Recently, mechanical techniques of laparoscopy sterilization have been developed in order to avoid the potential hazards associated with electrocautery. Two such techniques use spring-loaded clips or silicone rubberbands. The effectiveness is equivalent to that of the electrodes with potentially fewer complications. Long term statistics are still needed in assessing the long term effectiveness.[80,81]

Advantages to laparoscopy tubal ligation are that it

1. can be done under local or general anesthesia
2. can be done on an outpatient basis
3. leaves no scar
4. has low morbidity
5. is low cost

Disadvantages are

1. requires expensive instruments
2. complications few but can be significant (e.g., viscus perforation, perforation of major vessel, hemorrhage, thermal injury to bowel, cardiac arrhythmias or infection) (Table 7-21)
3. failure rate and ability to reanastomose not fully evaluated.

Table 7-21. Summary of complications of surgical laparoscopy.

	Number	*Rate/1000 cases*
Total laparoscopies reported	12,182	
Minor complications	130	10.7
Major complications	83	6.8
Anesthesia	9	0.75
Insufflation	9	0.75
Surgical trauma	38	3.2
Electrocoagulation	16	1.3
Cystitis or P.I.D.	10	0.8
Deaths	3	0.25

Data from Hulka, J. F., and Soderstrom, R. M.: Complication Committee of the American Association of Gynecologic Laparoscopists. J. Reprod. Med. 10:301, 1974.

Problem: Patient would like to know if her tubes can grow back together.

Answer: Recanalization can occur but it is uncommon. Pregnancy rates following tubal ligations are generally less than 1 percent. Patients should be informed of this potential of failure.

Problem: The patient asks if her tubes can be repaired.

Answer: Yes, under certain circumstances. Tubal reanastomosis can be done if the fimbria are present and if the removal did not include too large a segment of tube. The patency rate following this procedure is as high as 90 percent with a 40 to 50 percent pregnancy rate. The ectopic pregnancy rate may be increased. The rate of success with tubal reimplantation is discouraging.

The use of conventional unipolar, high-voltage electrocoagulation, widely used in recent years, has been associated with more extensive destruction than observed at the time the procedure was done. The surgeon, therefore, would be far less able to repair these tubes. The use of the bipolar coagulator, clips, bands or even the minilaparotomy may be preferable in the younger patient. However, reversibility should never be assured.[82]

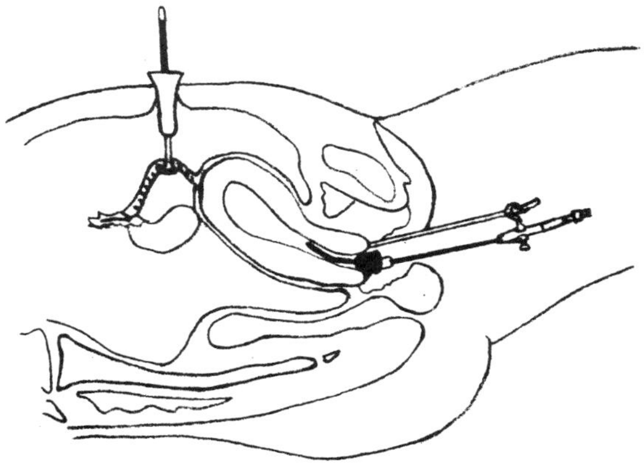

FIGURE 7-49. Technique of female sterilization by minilaparotomy in which tenaculum and Rubins' cannula are used to manipulate the uterus as in laparoscopy. A proctoscope is placed into the incision for visualization of the tubes; a forceps is inserted through the proctoscope to grasp the tube. (From Wortman, J.: *Female sterilization by mini-laparotomy*. Population Report, Series C, No. 5, November 1974, p. 57, with permission.)

Suprapubic Minilaparotomy. Interval female sterilization can be done by a suprapubic minilaparotomy incision with concomitant uterine elevation (Fig. 7-49). This procedure can be performed under local anesthesia and occasionally on an outpatient basis.[83]

Hysterectomy for Sterilization. The three types of hysterectomy are vaginal, total abdominal and cesarean (done concomitantly with cesarean section).

Advantages are

1. In the presence of pelvic pathology, e.g., symptomatic uterine decensus or myomata, hysterectomy can eliminate a source of future problems.
2. No further menstruation.
3. Virtually 100 percent effective.
4. Eliminates risk of uterine, albeit not vulvar, vaginal, or ovarian, cancer in group of susceptible women with poor follow-up potential.
5. Avoids the need of subsequent gynecologic surgery, i.e., 36 percent of patients having puerperal tubal ligation required subsequent gynecologic surgery.[84]

Disadvantages include

1. irreversibility
2. increased morbidity
3. prolonged hospitalization
4. no further menstruation
5. more costly than other methods
6. potential for psychological problems

Hysteroscopy. The tubal ostia can be physically obstructed with cautery, sclerosing chemicals or even silicone plugs, via a transcervical intrauterine approach utilizing the hysteroscope.

Advantages to this method are

1. low morbidity
2. can be done as an outpatient
3. occasionally can be done under local anesthesia

Disadvantages include

1. currently unacceptable high failure rate (may relate to enormous regenerative powers of the oviduct)
2. ability to repair doubtful
3. potential for perforation, bowel injury, infection or embolization

Chemical Occlusive Agents. A variety of chemical agents have been instilled transcervically for the purpose of occluding the fallopian tubes. These include ethanol, formaldehyde and silver nitrate. The failure rates and morbidity have so far been unacceptable. Current investigation with other agents, specifically quinicrine solutions, may offer better promise.

Sequelae to Female Sterilization. *Emotional.* The majority of patients report satisfaction with the procedure and a variable change in libido.

Psychological problems may be avoided with proper patient selection and counselling. Motivating stresses such as pregnancy should be carefully assessed to avoid later regrets. The effect of the following on coital enjoyment and total responsiveness should be understood prior to undertaking permanent contraception, specifically strong religious overtones; marital instability; fear of loss of mate or ability to reproduce; and misinterpretation about reversibility.

Physical. An increased occurrence of dysfunctional uterine bleeding following tubal ligation has been reported. Etiological factors are ambiguous. There are occasional complaints of pelvic pain, dysmenorrhea and frigidity.

Male Sterilization

Vasectomy. Vasectomy is an increasingly popular, effective, simple, safe and inexpensive method of fertility regulation. It involves removal of a segment of the vas deferens (Fig. 7-50).

Problem: The patient asks how soon he will be protected after this operation.

Answer: Sixty-five percent of patients will be sperm-free after 12 ejaculations, ninety-eight percent after 24 ejaculations, and it is virtually absolute after 36 ejaculations. The patient must be advised to use a contraceptive until

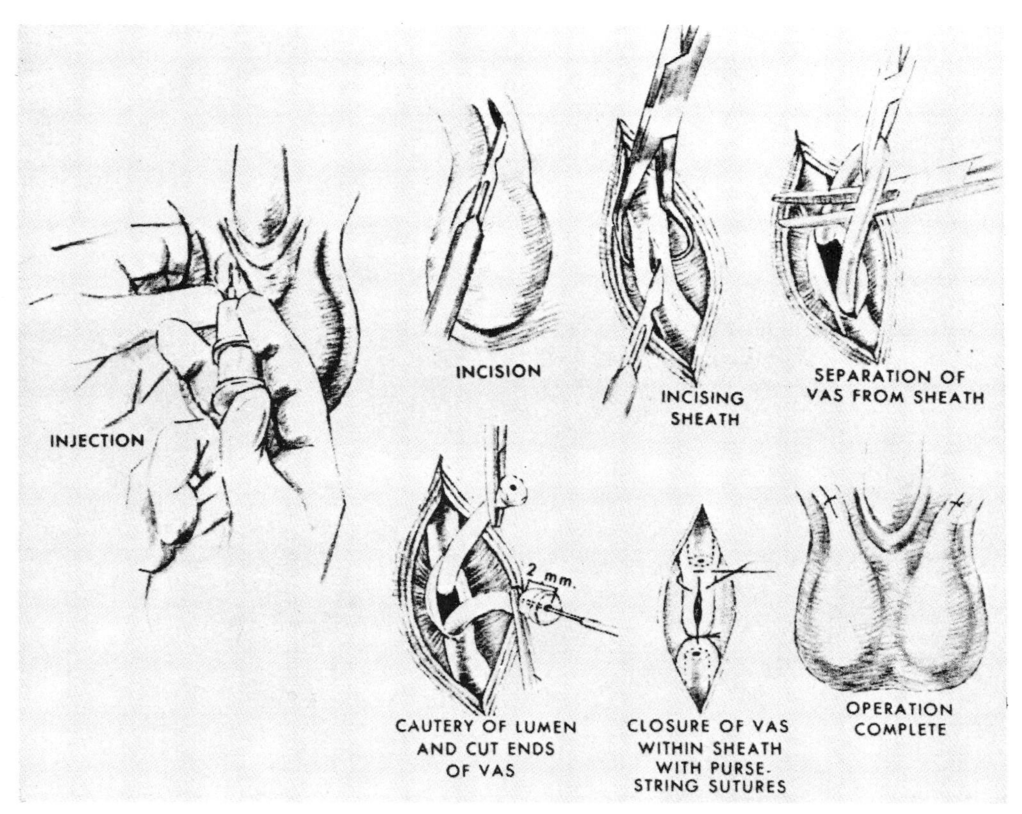

FIGURE 7-50. Technique of vasectomy. (From Hackett, R. E., and Waterhouse, K.: *Vasectomy—reviewed.* Am. J. Obstet. Gynecol. 116:438, 1973, with permission.)

there are two consecutive azospermic semen analyses. The patient should have follow-up semen analysis in 6 and 12 months. *At no time* should the individual be told that it is *impossible* to impregnate his mate.

Complications. The incidence of complications is low and all are minor (Fig. 7-51). Temporary use of a suspensory better assures greater comfort and reduced morbidity. Most complaints are short term and minor.

Operative complications may be treated in the following manner:

1. pain—with analgesics
2. edema—with elevation and ice packs
3. infection—with prophylactic antibiotics
4. nonspecific epididymitis—with aspirin, heat and suspension
5. hematoma—with ice and elevation; occasionally requires evacuation

The incidence of recanalization is low (0 to 6 percent) (Table 7-22).

There may be sperm granuloma as a result of extravasated sperm. Generally the condition is asymptomatic but occasionally it is painful. Many subside spontaneously after conservative treatment with ice bags, rest and anti-inflammatory agents. Occasionally, it requires excision. The incidence may be reduced by fulguration and/or the use of clips.

Sperm antibodies are present in 50 percent of men following vasectomy, in 2 percent of normal men and 5 percent in infertile men. They occur in direct proportion to the incidence of granuloma formation. The significance is unknown but studies to date support its inconse-

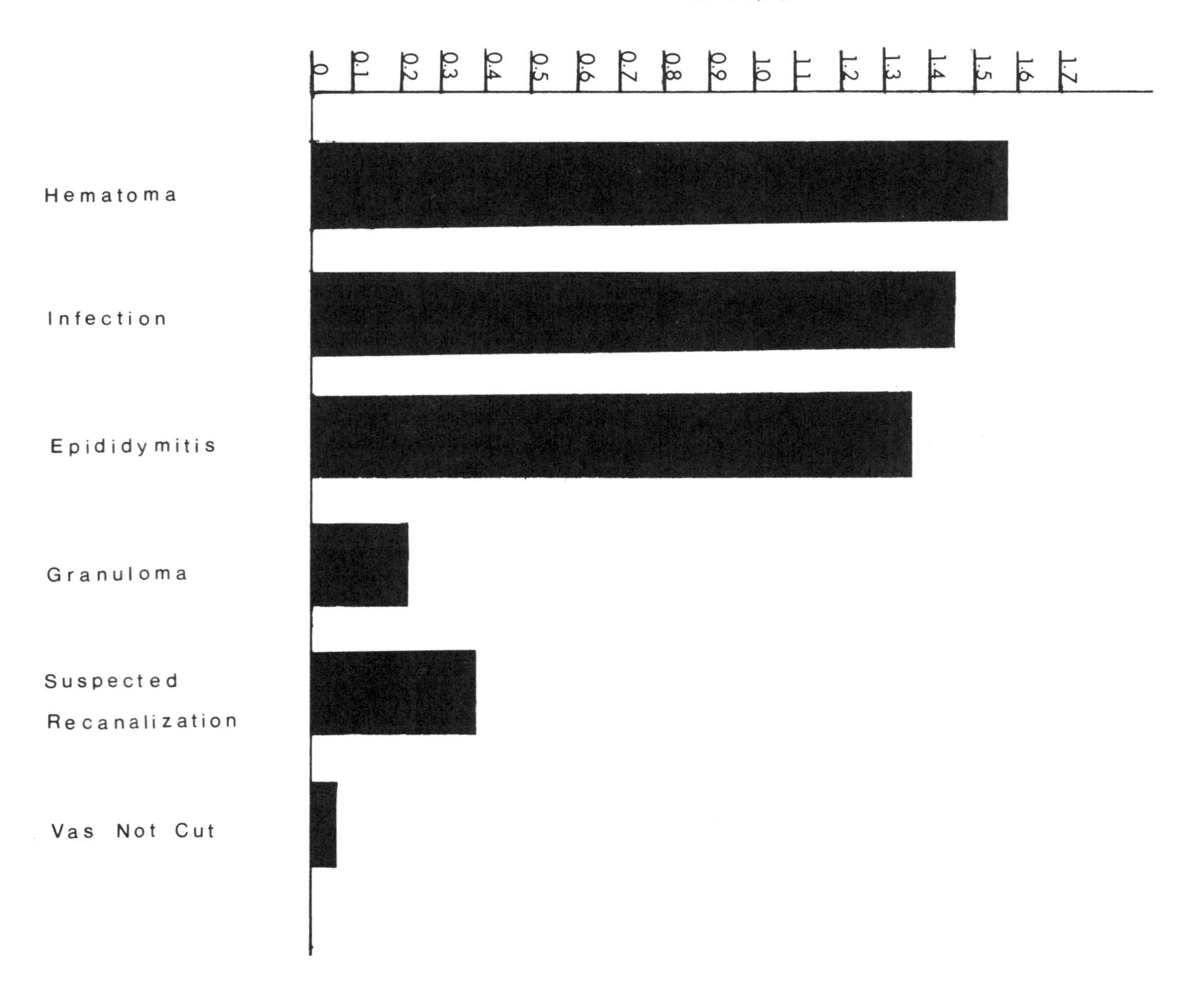

FIGURE 7-51. Vasectomy complications, selected studies, 1969–1974. (From Wortman, J.: *Vasectomy—what are the problems?* Population Report, Series D, No. 2, January 1975, p. 29, with permission.)

Table 7-22. Reported failure of vasectomy for three specified countries.

Country	*Number of failures*	*Failures per 100 cases*
India	6	0.5
Pakistan	2	0.3
Pakistan	3	2.2
United States	6	2.0
United States	3	4.1
All studies	20	0.9

Adapted from Presser, H. B.: *Voluntary sterilization: A world view.* Reports on Pop./Fam. Plann., No. 5, July, 1970, The Population Council, New York.

quence. The level may relate to failures of fertility restoration following vasovasotomy.

The average time to resumption of coitus is two weeks. Fifteen percent of the patients will have pain on their first coital exposure.

There are no or little psychologic problems involved.[85] Ninety-nine percent of patients studied had no regrets, 73 percent noted an increase in sexual pleasure and only 1.5 percent noted a decrease in sexual pleasure.

The wife's opinion of how vasectomy will affect her husband's potency is related to her level of education. About 15 percent of wives feel that this will interfere with their husband's ability to perform sexually.

General Problems and Answers. Problem: Patient would like to know what effects vasectomy will have on his "male physiology."

Answer: There is no direct measurable effect of ejaculation and semen volume; functions of the prostate seminal vesicles and urethral glands are unchanged. (2) There is no change in plasma testosterone and seminal fructose levels (semen acid phosphatase is increased); gonadotropins are unchanged. (3) Postvasectomy testicular biopsy shows normal spermatogenesis.

Problem: Is this operation reversible?

Answer: Yes. Patency rates by surgical reanastomosis are reported to be from 45 to 90 percent. Pregnancy rates are lower.

Problem: Can mechanical devices be used?

Answer: Intravas plugs, removable clips, reversible valves are being developed. Conclusion cannot be drawn at this time; however, preliminary data are not encouraging.

Problem: Are there any other alternatives to reversibility?

Answer: Yes. Sperm banks are available for those individuals whose product would give a good yield on thawing although the duration of preservation may be limited. There is a 30 to 60 percent recovery of live motile sperm on thawing. There is as yet no documentation of congenital anomalies resulting from this technique. However, adequate experience is lacking. Donor insemination also may be an acceptable alternative.[86,87,88]

Summary

Effectiveness: extremely high.

Advantages: no coital involvement, motivation or equipment. It is permanent.

Disadvantages: Transient inconvenience. Permanent contraception. In our culture many young couples will note personal tragedy (e.g., death of a child) or a change in life patterns (e.g., divorce and remarriage) which may influence their ideas on reproduction.

Risk of failure: low.

Alleged medical risk: small surgical risk; sometimes there is a psychological reaction.

FUTURE CONSIDERATIONS

No immediate prospects are visible in the near future. However, at the laboratory level, the following areas are being paid considerable attention.

1. Immunologic control—antibodies interfering with gonadotrophin activity (in particular, beta-subunit of HCG-tetanus toxoid immunization[89]), trophoblastic development, fertilization process, spermatozoa viability and releasing hormone functioning.
2. Retardation of epididymal spermatozoa maturation (e.g., antiandrogen-cyproterone acetate). No presently available compounds are acceptable to the stringent FDA screening.
3. Steroidal suppression of spermatozoa production (as above)—one that would be re-

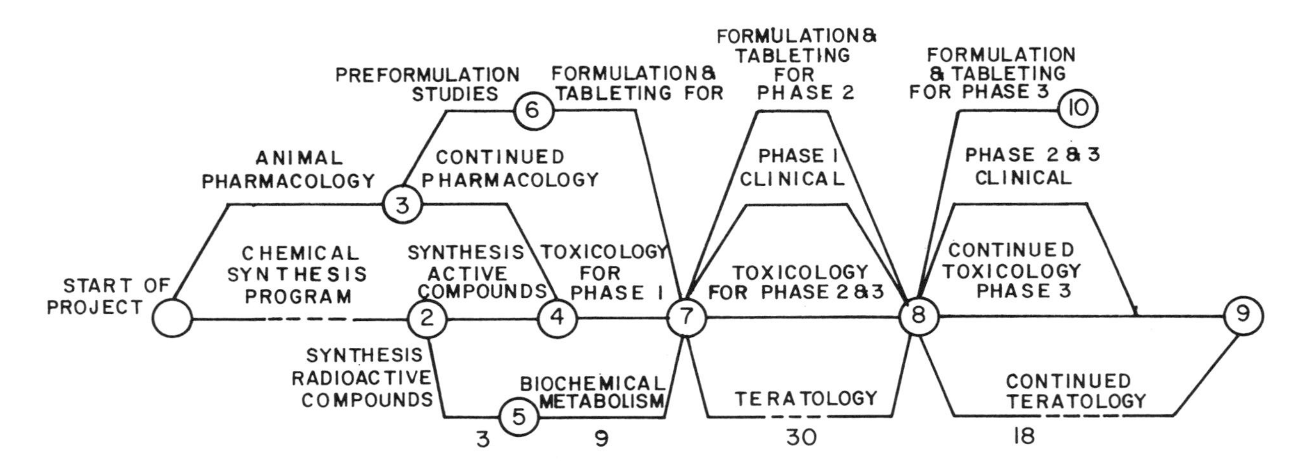

FIGURE 7-52. Basic critical path map of luteolytic or abortifacient agent. (Adapted from Christie, G. A.: *Rate limiting factors in the development of new contraceptive methods.* In Potts, M., and Wood, C. (eds.): *New Concepts in Contraception,* University Park Press, Baltimore, 1972, p. 150.)

Table 7-23. Requirements for clinical trial permission.

Documentation (U.S.A.)	*U.S.A./Canada*	*U. K.*/South Africa and New Zealand†*
IND Phase 1 (a few subjects for up to 10 days)	90-day studies in rats, dogs, monkeys	90-day studies in 2 species—1 rodent, 1 non-rodent (and possibly teratology and fertility studies)
IND Phase 2 (50 subjects for 3 menstrual cycles	1-year studies in rats, dogs, monkeys	6-month study in 2 species—1 rodent, 1 non-rodent. Teratology studies in 2 species
IND Phase 3 (large scale clinical trial)	2-year studies in rats, dogs, and monkeys. Also start of 7-year studies in dogs and ten-year studies in monkeys. Reproduction and teratology studies in 2 species	No specific details, but studies for 2 years in rats and 80 weeks in mice are generally commenced at this time
NDA (marketing)	No further requirements but up-to-date progress reports of at least 2 years duration on long term dog and monkey studies	Completion of 2-year rat and 80-week mouse studies. Metabolic studies—absorption, distribution, excretion

Adapted from Christie, G. A.: *Rate limiting factors in the development of new contraceptive methods.* In Potts, M., and Wood, C. (eds.): *New Concepts in Contraception.* University Park Press, Baltimore, 1972.

*Unlike the United States, the United Kingdom lays down no absolute requirements and judges each case on its merits.
†South Africa and New Zealand requirements follow the United Kingdom requirements except (1) they are generally less stringent regarding metabolic data and (2) they are more stringent regarding efficacy.

versible and without depression of serum testosterone.

4. Luteolytic agents—specifically affecting the ovarian steroidogenic enzymes. To date the adrenal system is equally affected by the compounds screened.

With the relative decrease in the amount of governmental funding and particularly with the increase in the governmental restrictions regarding regulations controlling drug development and human experimentation, there is some reluctance in the scientific community and the drug industry to develop newer and better methods of contraception. Figure 7-52 and Table 7-23 better indicate this complexity.

While the research is still going on, the need has intensified for the development of a more universally-acceptable contraceptive. That is one which is highly effective, easily used, independent of the coital act, easily reversible, nontoxic, inexpensive and able to meet individual needs of such as religious beliefs, cultural mores, socioeconomic background, intellectual motivation and emotional needs. One can readily see that no one specific agent could serve as an ideal contraceptive; hence the need for many approaches to meet the needs of varied individuals. Flexibility on the part of the physician is essential in order to accommodate the differences within a particular society.

In assessing the value of a particular contraceptive as well as the particular patient's needs, the health care provider must always weigh the relative risks against the advantages of the particular method selected for the specific person (Fig. 7-53). Moreover, confirmed sustained motivation of contraceptive adherence should be given particular attention.

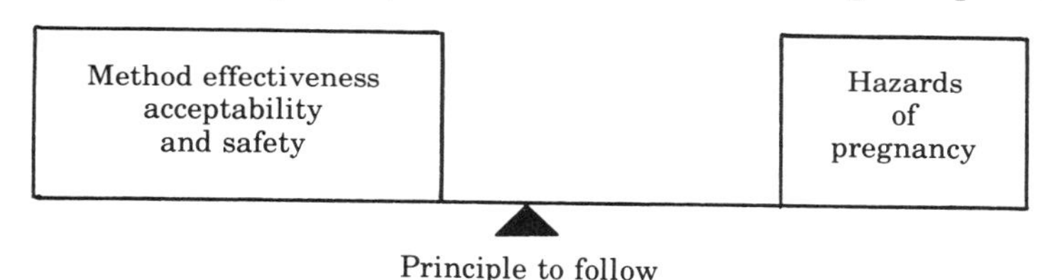

FIGURE 7-53. Scale of relative benefits.

SUMMARY

The past decade has witnessed the acceptance of family planning services as an entrant into the health care system for women. The ability of fertility control to enhance not only the physical well-being but also the quality of each person's life is now recognized. Unfortunately, the ideal method for contraception is missing, but research continues for a method which will be highly effective, easily used, independent of the coital act, easily reversible, nontoxic, inexpensive and able to meet individual desires such as religious beliefs, cultural mores, socioeconomic background, intellectual motivation and emotional needs. In the meantime, however, there are a variety of methods from which to choose. The health care provider must understand the effectiveness, advantages, disadvantages, risk of failure and alleged medical risks of each method, especially with regard to the needs of the individual patient.

QUESTIONS

1. In what ways does contraception contribute to health care?
2. What are other advantages of efficient contraception?
3. Should the physician take an active or passive role in initiating contraception? Why?
4. Are family planning services essential for a reduction in fertility? Discuss.
5. Explain the differences between the theoretical effectiveness of a contraceptive method and its actual use effectiveness.
6. What are the important determinants in prescribing a contraceptive modality?

REFERENCES

1. Jaffe, F. S.: *What private physicians do about family planning*. Comtemp. Obstet. Gynecol. 4:39, 1974.
2. Ibid.
3. Nash, E. M., and Louden, L. M.: *The premarital medical examination and the Carolina population center*. J.A.M.A. 210:2365, 1969.
4. Edwards, E. M.: *A study of contraceptive practices in a selected group of urban negro mothers in Baltimore*. Am. J. Public Health 58:263, 1968.
5. Chang-Silva, A. W., Mudd, E. H., and García, C.R.: *Psychosexual response and attitudes toward family planning*. Obstet. Gynecol. 37:289, 1971.
6. Ryder, N. B.: *Time series of pill and IUD use: United States 1961–1970*. Stud. Fam. Plann. 3:233, 1972.
7. Rochat, R. W., Tyler, C. W., and Schoenbucher, A. K.: *Advances in Planned Parenthood*. Excerpta Medica, New York, Vol. 6, 1971.
8. Okada, L. M.: *Use of matched pairs in evaluation of a birth control program*. Public Health Report 84:445, 1969.
9. Sivin, I.: *International postpartum family planning program*. Stud. Fam. Plann. 2:248, 1971.
10. Ravenholt, R. T., and Chao, J.: Pop. Report, Series J, No. 2, 1971, p. 21.
11. Ryder, N. B.: op. cit.
12. Nortman, D.: *Population and family planning programs: a factbook*. Reports on Population/Family Planning, No. 2 (ed. 5), September 1973.
13. Westoff, C. F.: *The modernization of U.S. contraceptive practice*. Fam. Plann. Perspect. 4:9, 1972.
14. Freedman, R., and Berelson, B.: *The record of family planning programs*. Stud. Fam. Plann. 7:1, 1976.
15. Ibid, p. 11.
16. Goldzieher, J. W.: *How the oral contraceptive came to be developed*. J.A.M.A. October 1974.
17. Ryder, N. B.: *Contraceptive failure in the U. S.* Fam. Plann. Perspect. 5:133, 1973.
18. Tietze, C.: *Probability of pregnancy resulting from a single unprotected coitus*. Fertil. Steril. 11:485, 1960.
19. Westoff, C. F., and Bumpass, L.: *The revolution in birth control practices of U. S. Roman Catholics*. Science 179:41, 1973.
20. Forrest, J. E.: *Postpartum services in family planning: findings to date*. Report on Population/Family Planning 8:2, 1971.
21. Van Ginneken, J. K.: *Prolonged breastfeeding as a birth spacing method*. Stud. Fam. Plann. 5:177, 1974.
22. Perez, A., et al.: *The first ovulation after childbirth: the effect of breast feeding*. Am. J. Obstet. Gynecol. 114:1041, 1972.
23. Ryder, N. B.: *Contraceptive failure . . .*, op. cit.
24. Westoff, C. F., and Bumpass, L.: *Oral contraception, coital frequency, and the time required to conceive*. Soc. Biol. 16:1, 1969.
25. Wilkinson, E., and DuFour, D. R.: *Pathogenesis of microglandular hyperplasia of the cervix uteri*. Obstet. Gynecol. 47:189, 1976.
26. Inman, W. H. W., and Vessey, M. P.: *Investigation of deaths from pulmonary, coronary, and cerebral thrombosis and embolism in women of child-bearing age*. Br. Med. J. 2:193, 1968.
27. Vessey, M. P., and Doll, R.: *Investigation of relation between use of oral contraceptives and thromboembolic disease: a further report*. Br. Med. J. 2:651, 1969.
28. Sartwell, P. E., et al.: *Thromboembolism and oral contraceptives: an epidemiologic case-control study 1, 2*. Am. J. Epidem. 90:365, 1969.
29. Royal College of General Practitioners. Pitman Publishers, Inc., New York, 1974.
30. Ibid.
31. Vessey, M. P., et al.: *Post-operative thromboembolism and the use of oral contraceptives*. Br. Med. J. 3:123, 1970.
32. Green, G. R., and Sartwell, P. E.: *Oral contraceptive use in patients with thromboembolism following surgery trauma or infection*. Am. J. Pub. Health 62: 1972.
33. Weir, R. J., et al.: *Blood pressure in women after one year of contraception*. Lancet 1:467, 1971.
34. Fisch, I. R.: *Oral contraception, pregnancy, and blood pressure*. J.A.M.A. 222:1507, 1972.
35. Royal College of General Practitioners, op. cit.
36. Spellacy, W. N., and Birk, S. A.: *The effects of mechanical and steroid contraceptive methods on the blood pressure in hypertensive women*. Fertil. Steril. 25:467, 1974.
37. Mann, J. Z., and Inman, W. H. W.: *Oral contraceptives and death from myocardial infarction*. Br. Med. J. 2:25, 1975.
38. Mann, J. Z., et al.: *Myocardial infarction in young women with special reference to oral contraceptive practice*. Br. Med. J. 241–245, 1975.
39. Shapiro, S.: *Oral contraceptives and myocardial infarction*. N. Engl. J. Med. 293:195, 1975.
40. Ory, H., Cole, P., McMahon, B., and Hoover, R.: *Oral contraceptives and reduced risk of benign breast disease*. N. Engl. J. Med. 294:419, 1976.
41. Schaffner, F.: *The effect of oral contraceptives on the liver*. J.A.M.A. 198:1019, 1966.
42. *Boston collaborative drug surveillance program*. Lancet 1:1399, 1973.
43. Royal College of General Practitioners, op. cit.
44. Bennion, L. J., et al.: *Effects of oral contraceptives on the gallbladder bile of normal women*. N. Engl. J. Med. 294:189, 1976.
45. Baum, J., et al.: *Possible association between benign hematomas and oral contraceptives*. Lancet 2:926, 1973.
46. O'Sullivan, J. P., and Wilding, R. P.: *Hematomas in patients on oral contraceptives*. Br. Med. J. 3:7, 1974.
47. Edmonson, H. A., Henderson, B., and Benton, B.: *Liver cell adenomas associated with the use of oral contraceptives*. N. Engl. J. Med. 294:470, 1976.
48. García, C. R.: *Medical and metabolic effects of oral contraceptives*. Clin. Obstet. Gynecol. 2:674, 1968.
49. Carr, D. H.: *Chromosome studies in selected spontaneous abortions: 1. conception after oral contracep-*

tives. Can. Med. Assoc. J. 103:343, 1970.

50. Rice-Wray, E.: *Pregnancy and progeny after hormone contraceptives—genetic studies*. J. Reprod. Med. 6:101, 1971.
51. Nora, A. H., and Nora, J. J.: *A syndrome of multiple congenital anomalies associated with teratogenic exposure*. Arch. Environ. Health 30:17, 1975.
52. Janerich, D. T., Piper, J. M., and Glebatis, D. M.: *Oral contraceptives and congenital limb-reduction defects*. N. Engl. J. Med. 291:697, 1974.
53. Lanman, J. T., and Jain, A.: *Association of oral contraceptives and congenital limb-reduction defects*. Population Council, October, 1974, p. 4.
54. Littlefield, L. G., and Mailhes, J. B.: *Comparison of chromosome breakages in lymphocytes and fibroblasts from control women and women taking oral contraceptives*. Fertil. Steril. 26:828, 1975.
55. Editorial. Br. Med. J. 4:59, 1972.
56. Golditch, I. M.: *Postcontraceptive amenorrhea*. Obstet. Gynecol. 39:903, 1972.
57. Westoff, C. F., and Bumpass, L.: op. cit.
58. Larson-Cohn, J.: *Oral contraceptives and vitamins: a review*. Am. J. Obstet. Gynecol. 121:84, 1975.
59. Huber, D. H., and Huber, S. C.: *Screening oral dispensable?* Presented at the Association of Planned Parenthood Physicians, Los Angeles, April 18, 1975.
60. Statement by the Central Medical Committee of the International Planned Parenthood Federation, April 1973. Reprinted from Piotrow, P.: *Oral contraceptives—50 million users*. Population Report, Series A, No. 1, 1974, with permission.
61. Moghissi, K. S., et al.: *Contraceptive mechanism of microdose norethindrone*. Obstet. Gynecol. 41:585, 1973.
62. Laurie, R. E., and Korba, U. D.: *Fertility control with continuous microdose norgestrel*. J. Reprod. Med. 8:165, 1972.
63. Scutchfield, F. C., et al.: *Medroxyprogesterone acetate as an injectable female contraceptive*. Contraception 3:21, 1971.
64. Schwallie, P. C., and Assento, J. R.: *Contraceptive use-efficacy study utilizing Depo-Provera administered as an injection every six months*. Contraception 6:315, 1972.
65. Schwallie, P. C.: *Experience with Depo-Provera as an injectable contraceptive*. J. Reprod. Med. 13:113, 1974.
66. Rosenfield, A. G.: *Injectable long-acting progestogen contraception: a neglected modality*. Am. J. Obstet. Gynecol. 120:537, 1974.
67. Zañartu, J., et al.: *Effect of a long-acting contraceptive progestogen on lactation*. Obstet. Gynecol. 47:174, 1976.
68. Kuchera, L. K.: *Post coital contraception with diethylstilbestrol*. J.A.M.A. 218:562, 1971.
69. Morris, J. M., and Van Wagenen, G.: *Interception: the use of postovulatory estrogens to prevent conception*. Am. J. Obstet. Gynecol. 115:101, 1973.
70. Rosenfeld, D. L., et al.: *Medical, psychologic, and social factors in morning-after pill utilization*. Advances in Planned Parenthood 11:19, 1976.
71. Tietze, C.: *Evaluation of intrauterine devices: ninth progress report of the Cooperative Statistical Program*. Stud. Fam. Plann. 1:1, 1970.
72. Targum, S. D., and Wright, N. H.: *Association of the intrauterine device and pelvic inflammatory disease: a retrospective pilot study*. Am. J. Epidemiol. 100:262, 1974.
73. Mishell, D. R.: *Assessing the intrauterine device*. Fam. Plann. Perspect. 7:103, 1975.
74. Alvior, G. T.: *Pregnancy outcome with removal of intrauterine device*. Obstet. Gynecol. 41:894, 1973.
75. Christian, C. D.: *Maternal deaths associated with an intrauterine device*. Am. J. Obstet. Gynecol. 119:441, 1974.
76. Rosenfield, A. G., and Castadot, R. G.: *Early postpartum and immediate postabortion intrauterine contraceptive device insertion*. Am. J. Obstet. Gynecol. 118:1104, 1974.
77. Oster, G., and Salgo, M. P.: *The copper intrauterine device and its mode of action*. N. Engl. J. Med. 293:432, 1975.
78. Tatum, H. J.: Am. J. Obstet. Gynecol. 117:603, 1973.
79. Gonzales, B.: *Estimate of numbers of voluntary sterilizations performed*. Press release of Association for Voluntary Sterilization, New York, 1975, appeared in Editorial: Fam. Plann. Perspect. 7:274, 1975.
80. Hulka, J. F., et al.: *Sterilization with the spring-loaded clip*. Fertil. Steril. 26:1122, 1975.
81. Yoon, I. B., and King, T. M.: *Sterilization with the falope ring* in Sciarra, J. S. (ed.): *Advances in Female Sterilization Techniques*. Harper and Row Publishers, Inc., New York, 1976 (in press).
82. Rosenfeld, D. L., and García, C. R.: *Laparoscopy prior to tubal reanastomosis*. J. Reprod., 1976 (in press).
83. Osathanondh, Vitoon: Contraception 10:251, 1974.
84. Haynes, D. M., and Wolfe, W. M.: *Tubal sterilization in an indigent population report of fourteen years' experience*. Am. J. Obstet. Gynecol. 106:1044, 1970.
85. Presser, H. B., and Bumpass, L. L.: *The acceptability of contraceptive sterilization among U. S. couples: 1970*. Fam. Plann. Perspect. 4:18, 1972.
86. Hackett, R. E., and Waterhouse, K.: *Vasectomy—reviewed*. Am. J. Obstet. Gynecol. 116:438, 1973.
87. Saberro, A. J., and Kohli, K. L.: *Vasectomy*. Stud. Fam. Plann. 5:42, 1974.
88. Wortman, J.: *Vasectomy—what are the problems?* Population Reports, Series D, No. 2, p. 25, 1975.
89. Talwar, G. P., et al.: *Processing of the preparations of beta-subunit of human chorionic gonadotropin for minimization of cross-reactivity with human luteinizing hormone*. Contraception 13:131, 1976.

BIBLIOGRAPHY

Belsky, R.: *Vaginal contraceptives, a time for reappraisal*. Population Report, Series H, No. 3, January 1975.

Dalsimer, I. A., Piotrow, P. T., and Dumm, J. J.: *Barrier methods*. Population Report, Series H, No. 1, December 1973.

García, C. R.: *Ten years' evaluation of human pharmacology of the oral contraceptives*. Clinical Trails J. January 1968.

Goldzieher, J. W.: *An assessment of the hazards and metabolic alterations attributed to oral contraceptives*. Contraception 1:409, 1970.

Hackett, R. E., and Waterhouse, K.: *Vasectomy— reviewed*. Am. J. Obstet. Gynecol. 116:438, 1973.

Huber, S. C., et al.: *IUDs reassessed—a decade of experience*. Population Report, Series B, No. 2, 1975.

Jaffe, F. S.: *What private physicians do about family planning*. Contemp. Obstet. Gynecol. 4:39, 1974.

Mastoianni, L.: *Rhythm: systematized chance-taking*. Fam. Plann. Perspect. 6:209, 1974.

Mishell, D. R.: *Assessing the intrauterine device*. Fam. Plann. Perspect. 7:103, 1975.

Piotrow, P. T., et al.: *Oral contraceptives*. Population Reports, Series A, No. 2, March 1975.

Potts, M., and Wood, C. (eds.): *New Concepts in Contraception*. University Park Press, Baltimore, 1972.

Presser, H. B.: *Voluntary sterilization: a world view*. Reports on Population/Family Planning, 5:1, 1970.

Boss, J. A., et al.: *Findings from family planning research*. Reports on Population/Family Planning 12:1, 1972.

Ryder, N. B.: *Contraceptive failure in the U. S.* Fam. Plann. Perspect. 5:133, 1966.

Shepard, M. K.: *Female contraceptive sterilization*. Obstet. Gynecol. Survey 29:739, 1974.

Siegel, D., and Corfman, P.: *Epidemiological problems with studies of the safety of oral contraceptive*. J.A.M.A. 203:950, 1968.

Westoff, C. F.: *The modernization of U. S. contraceptive practice*. Fam. Plann. Perspect. 4:9, 1972.

8 Pregnancy Termination

The medical, ethical and social considerations of pregnancy termination have provoked enormous controversy. The purpose of this chapter is to direct the attention of the reader to some of these issues—legalization, advantages, disadvantages, risks, complications, psychological impact, patient characteristics and techniques.

Abortion has probably been the most widely resorted to and effective form of birth limitation throughout the history of mankind. It was condemned by Hippocrates and championed by Aristotle. Today more than 70 percent of the world's population live in countries where legal abortion is available for "health" reasons, albeit other than for strictly classical medical reasons.[1] Curiously the surreptitious use of abortion is universal in all countries with restrictive abortion laws. For example, in European countries the estimate of the abortion to live birth ratio is from 0.8/1 to 1.3/1. The actual incidence of abortion is probably underestimated (Table 8-1, Fig. 8-1).

Table 8-1. Percent of female population of reproductive age who have had abortion.

Country	*Percent*
Chile	27.2
Hungary	29.0
Japan	32.2
S. Korea	14.0

Data from Ross, J. A., et al.: *Findings from family planning research*. Reports on Population/Family Planning 12:35, 1972.

ABORTION LEGISLATION

The legality of the abortion itself has meant much for the patient concerned and for the determination of the type of procedure and the safety of that service rendered, but legality probably has had much less bearing on the overall abortion rate than the figures imply. For example, Italy, where abortion had been illegal, was estimated to have had one of the highest abortion rates in the world—1.3 abortions per live birth. Another example is in Peru where the civil code punishes the female who induces her own abortion with four years imprisonment and the abortionist with two to ten years. An abortion performed by a health professional results in a suspension of privileges for at least five years. Despite these laws the abortion/live birth ratio is estimated to be 46.5 percent.[2]

The attitudes of the patient and of society towards abortion have been quite divergent, ranging from fatalism to punishment to self-determination. Abortion legislation therefore embraces many philosophies, i.e.:

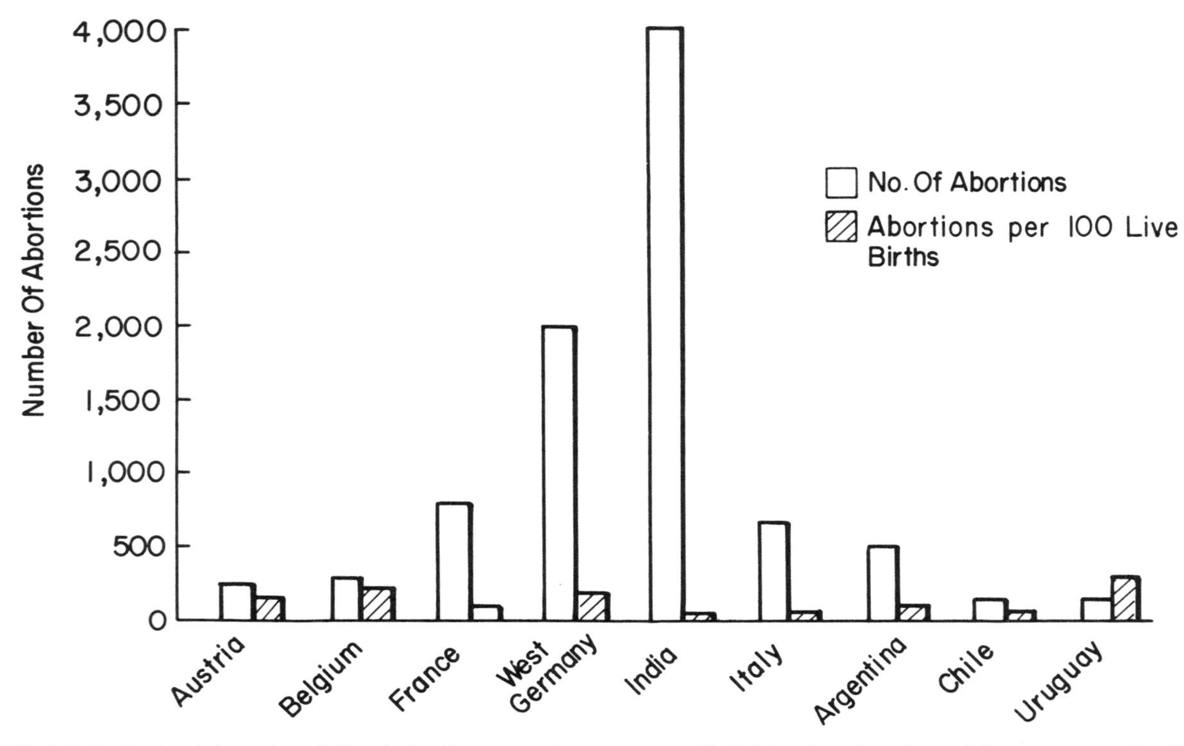

FIGURE 8-1. Estimated number of illegal abortions in various countries, 1967. (Based on data from af Geijerstam, G.: *Fertility control by induced abortion.* In Diczfausy, E., and Borlell, U. (eds): *Control of Human Fertility.* Nobel Symposium 15. Almquvist & Wilsell, Amsterdam, 1971, p. 218.)

1. No abortion in any circumstances.
2. Abortion only if the mother's immediate survival is affected.
3. Abortion only if the mother's longevity is affected.
4. Self-determination—woman's right to choose.

The attitudes of most nations have been shifting gradually in favor of supporting the concept of legality of abortion. Nonetheless, there has concomitantly been an increased resistance against compliance with and attempts to reverse the new law by anti-abortion forces, who, because their earlier efforts were unsuccessful, are now directing attention to the protection of the legal rights of the unborn fetus. This stand unfortunately totally disregards historical precedents as stated above.

Table 8-2 lists the present status of abortion throughout the world in mid-1975. Note that "legal authorization of elective abortion does not guarantee that abortion on request is actually available to all women who may want their pregnancies terminated. Lack of medical personnel and facilities or conservative attitudes among physicians and hospital administrators may effectively curtail access to abortion, especially for economically or socially deprived women."[3]

Abortion Legislation in the United States

Prior to the liberalization of American abortion legislation there were estimated nearly one million abortions annually in the United States with five to 10,000 abortion-related maternal deaths. Legislation on abortion had traditionally been the responsibility of each state. On January 22, 1973, the United States Su-

Table 8-2. Legal status of abortion by country and by grounds: mid-1975.

Country	Population mid-1975 estimate (in millions)	Illegal	Legal on specified grounds[1]					Legal (grounds not specified)
			Medical					
			Narrow (life)	Broad (health)	Eugenic (fetal)	Juridical (rape, incest, etc.)	Social and social-medical	
Algeria	16.8		x					
Argentina	25.4			x		x		
Australia	13.8			x				
(New South Wales)[2]				x			x	
(South Australia)[2]				x	x		x[3]	
Austria	7.5							x[4]
Bangladesh	73.7		x					
Belgium	9.8	x						
Bolivia	5.4	x						
Brazil	109.7		x			x		
Bulgaria[5]	8.8			x	x	x	x[6]	
Cameroon	6.4			x		x		
Canada	22.8			x				
Chile	10.3			x				
China, Peoples Republic of	822.8							x[4]
Colombia	25.9	x						
Costa Rica	2.0			x				
Cuba[7]	9.5			x	x	x		
Czechoslovakia	14.8			x	x	x	x[4]	
Denmark	5.0							x[4]
Dominican Republic	5.1	x						
Ecuador	7.1			x		x		
Egypt, Arab Republic of	37.5	x						
El Salvador	4.1			x	x	x		
Ethiopia	28.0			x				
Finland[8]	4.7			x	x	x	x[9]	
France	52.9							x[6]
German Democratic Republic	17.2							x[4]
German Federal Republic	61.9			x				
Ghana	9.9			x				
Great Britain	54.9			x	x		x[3]	
Greece	8.9			x		x		
Guatemala	6.1		x		x			
Haiti	4.6	x						
Honduras	3.0			x				
Hong Kong	4.2			x	x		x[3]	
Hungary[10]	10.5			x	x	x	x[4]	
India	613.2			x	x	x	x[3]	
Indonesia	136.0	x						
Iran	32.9		x					
Iraq	11.1		x					
Ireland, Northern	1.5		x					
Ireland, Republic of	3.1	x						
Israel	3.4			x				
Italy	55.0			x				
Ivory Coast	4.9		x					
Jamaica	2.0	x						
Japan	111.1			x	x	x	x[3,11]	

Table 8-2. Legal status of abortion by country and by grounds: mid-1975. (*Continued*)

Country	*Population mid-1975 estimate (in millions)*	*Illegal*	*Legal on specified grounds*[1] *Medical Narrow (life)*	*Medical Broad (health)*	*Eugenic (fetal)*	*Juridical (rape, incest, etc.)*	*Social and social-medical*	*Legal (grounds not specified)*
Jordan	2.7			x		x		
Kenya	13.3			x				
Khmer Republic	8.1		x					
Korea, Republic of	33.9			x	x	x		
Lebanon	2.9		x					
Liberia	1.7	x						
Malaysia	12.1		x					
Mexico	59.2		x			x		
Morocco	17.5			x				
Netherlands[12]	13.6		x					
New Zealand	3.0		x					
Nicaragua	2.3		x					
Nigeria	62.9		x					
Norway	4.0			x	x	x	x[4]	
Pakistan	70.6		x					
Panama	1.7	x						
Paraguay	2.6		x					
Peru	15.3			x				
Philippines	44.4	x						
Poland	33.8			x		x	x[4,11]	
Portugal	8.8	x						
Romania[13]	21.2			x	x	x	x[4]	
Senegal	4.4		x					
Sierra Leone	3.0			x				
Singapore	2.2							x[14]
South Africa	24.7			x	x	x		
Spain	35.4		x					
Sri Lanka	14.0		x					
Sudan	18.3		x					
Sweden	8.3							x[13]
Switzerland	6.5			x				
Syria	7.3			x				
Taiwan	16.0	x						
Thailand	42.1			x		x		
Trinidad and Tobago	1.0	x						
Tunisia	5.7							x[4]
Turkey	39.9			x	x			
Uganda	11.4			x				
United States	213.9							x[4]
USSR	255.0							x[3]
Uruguay[16]	3.1			x		x	x[4]	
Venezuela	12.2		x					
Vietnam, Democratic Republic of	23.8							x[17]
Vietnam, Republic of South	19.7		x					
Yugoslavia	21.3			x	x	x	x[4]	
Zaire	24.5	x						
Zambia	5.0			x	x		x[3]	

Reprinted by permission of The Population Council from Tietze, C., and Mursten, M. C.: *Induced abortion: 1975 factbook.* Reports on Population/Family Planning No. 14, December 1975, p. 8.

Table 8-2. Legal status of abortion by country and by grounds: mid-1975. (*Continued*)

Note: Table does not include countries under one million inhabitants and those for which information on legal status of abortion was not located.

[1]Abortion on medical and eugenic grounds is generally permitted prior to viability. Abortion on juridical grounds is generally permitted up to the same gestational period as abortion on social and social medical grounds.

[2]Population included in Australia.

[3]Prior to viability of fetus.

[4]During first trimester (three months or twelve weeks)

[5]On request for married women with two living children, unmarried women, and women over age 40 with one living child.

[6]During first ten weeks.

[7]Abortion on request available in government hospitals.

[8]On request for women over age 40.

[9]During first sixteen weeks.

[10]On request for unmarried women, for married women with three children, for some married women with two living children, and for married women over age 40.

[11]No formal authorization procedures required and abortion permitted in doctor's office, hence, abortion de facto available on request.

[12]Abortion on request openly available in nonprofit clinics.

[13]Abortion on request for women over age 40 and women having four or more children.

[14]During first 24 weeks.

[15]During first 18 weeks.

[16]Penalty may be waived when abortion is performed for reasons of serious economic difficulty.

[17]Gestational limitation not ascertained.

preme Court ruled favorably on the legality of a woman's right to choose abortion. They concluded:

1. During the first trimester it is the woman's right to select the physician to perform an abortion.
2. During the second trimester the state can intervene only if that procedure does not conform to safe medical practices.
3. Abortion can be prohibited in the final trimester if there is no threat to the mother's life or health.

Attitudes of Americans regarding abortion are depicted in Figures 8-2 and 8-3.

Following the liberalization of the abortion laws, done initially at the state level (e.g., Colorado, Hawaii, New York, Washington, Maryland), there has been a marked increase in the utilization of legal pregnancy termination in the United States (Fig. 8-4, Table 8-3).

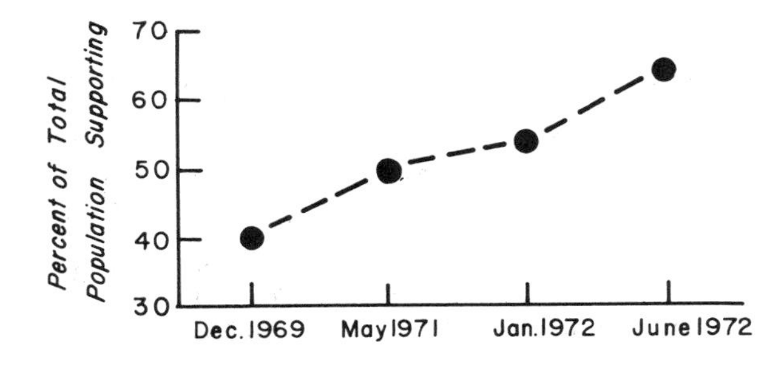

FIGURE 8-2. *Gallup Poll Issue:* Decision to have an abortion in the early months of pregnancy should be made by the woman (couple) and her (their) doctor.

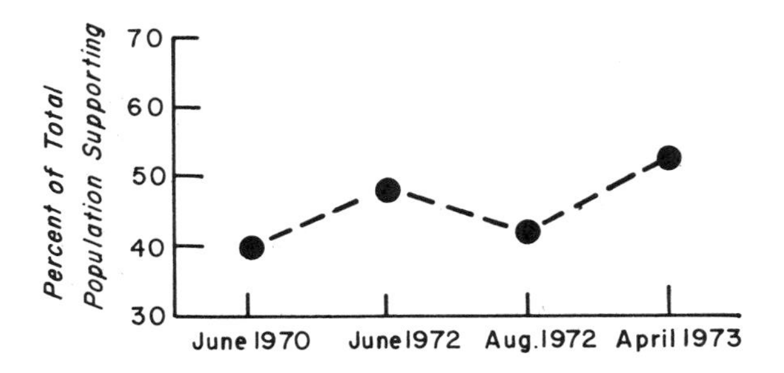

FIGURE 8-3. *Harris Survey Issue:* Those favoring legalized abortion up to four months of pregnancy.

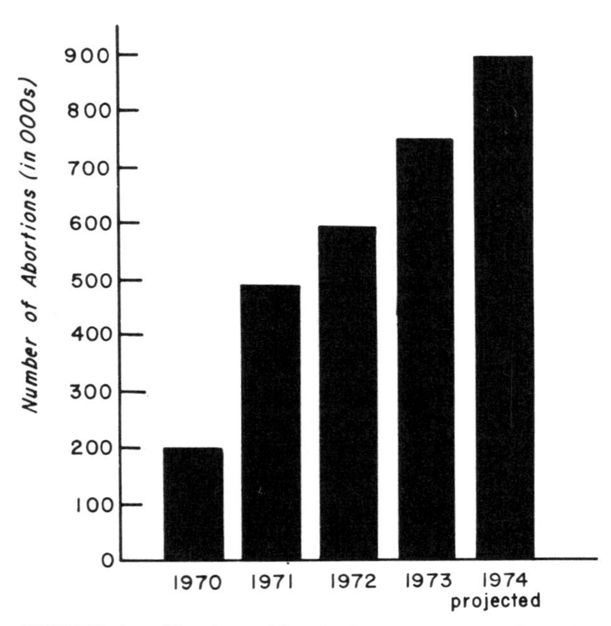

FIGURE 8-4. Number of legal abortions reported in the U.S., 1970 to 1973, and projected for 1974. (Adapted from Weinstock, E., et al.: *Legal abortions in the United States since the 1973 Supreme Court decision.* Fam. Plann. Perspect. 7:23, 1975.)

Table 8-3. Change in number of legal abortions in the United States.

Year	*No. of Abortions*
1969	1/167 births
1971	1/7 births
1973	1/4 births
1969	1/every 1,000 females ages 15–44
1971	11/every 1,000 females ages 15–44
1973	16/every 1,000 females ages 15–44

Data from Tyler, C. W.: *Abortion services and abortion seeking behavior in the United States.* In Osofsky, H. J., and Osofsky, J. D. (eds.): *The Abortion Experience.* Harper and Row Publishers, Inc., New York, 1973, p. 30.; and from Weinstock, E.: *Legal abortions in the United States since the 1973 Supreme Court decisions.* Fam. Plann. Perspect. 7:26, 1975.

ADVANTAGES OF LEGALIZED ABORTION

The high frequency of therapeutic abortion makes it one of the most commonly practiced surgical procedures second only to tonsillectomy. Recent estimates have ranged from 30 to 55 million abortions worldwide yearly. This corresponds to abortion rates from around 40 to 70 per 1000 women of reproductive age and to abortion ratios (the number of births in the denominator is the number occurring during a period of 12 months starting six months later than the period during which abortions were performed) of 260 to 450 per 1000 live births.[4] In the United States in 1973 there were nearly 750,000 legal abortions reported with an abortion ratio of 236 abortions per 1000 live births and an abortion rate of 3.5 per 1000 total population (16.5 per 1000 women ages 15 to 44).[5]

As with any other aspect of medical care, one must measure the relative merits and liabilities of that practice. Some will be discussed in this section.

One must consider the potential physical and mental damage to both the female in denying her an abortion and her undesired child. One must also reflect on the interpersonal relationship of the parents and their attitude toward an undesired child.

Easing of Population Pressures

The decline in a country's birth rate is accompanied by a rise in both contraceptive and abortion utilization.[6]

1. The most dramatic changes in fertility have been found in those countries of Eastern and Central Europe and in Japan where legalized abortion has been longstanding.
2. Associated with the legalization of abortion, Japan noted the most rapid decline (41 percent) ever in birth rate of any nation.
3. The decline in the general fertility rates in states with liberalized abortion laws (California and New York) in the year prior to the Supreme Court decision was nearly two times greater than the decline in fertility in the United States as a whole (6 percent).
4. In 1973 in New York City there was a 20 percent decrease in the number of births to teenagers and a 25 percent decrease in the teenage fertility rate (Fig. 8-5).

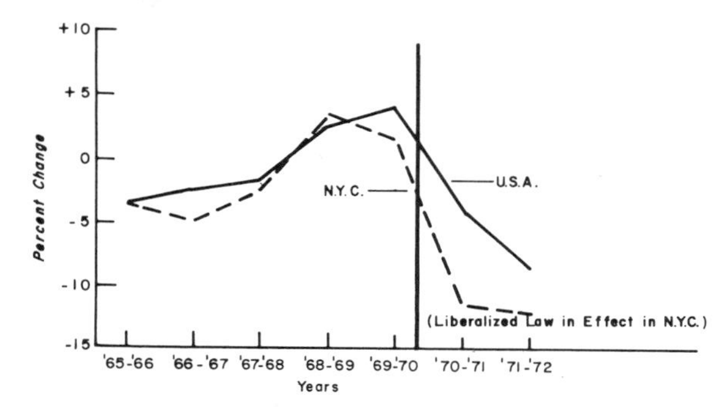

FIGURE 8-5. Annual change in number of live births in the U. S. and New York City, 1965 through 1972. (Adapted from Pakter, J., et al.: *A review of two years' experience in New York City with the liberalized abortion law*. In Osofsky, H. J., and Osofsky, J. D. (eds.): *The Abortion Experience*. Harper and Row Publishers, Inc., New York, 1973, p. 65.)

With time, as contraception availability and usage improves, there is an eventual decline in the number of abortions (Fig. 8-6). This has occurred in England and Japan.

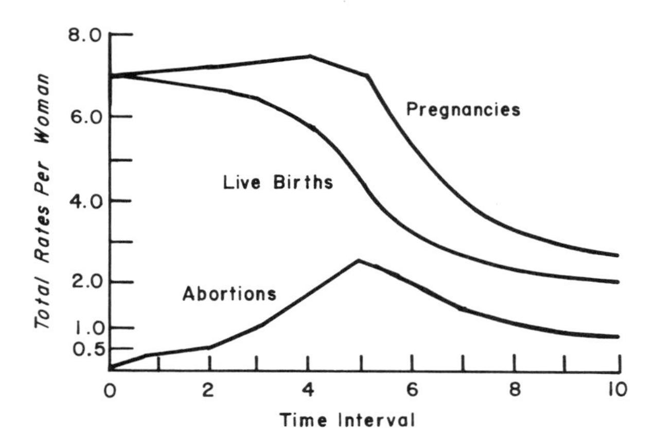

FIGURE 8-6. Transition from uncontrolled fertility to replacement level. (Reprinted by permission of The Population Council from Tietze, C., and Dawson, D. A.: *Induced abortion: a factbook*. Reports on Population/Family Planning 14:6, 1975.)

In those countries in which contraceptives are not readily available, the abortion rate remains at very high levels, e.g., Union of Soviet Socialist Republics.

Decline in Maternal Mortality

Prior to the liberalization of abortion laws, illegal abortion was the leading cause of maternal mortality (Figs. 8-7, 8-8 and 8-9).

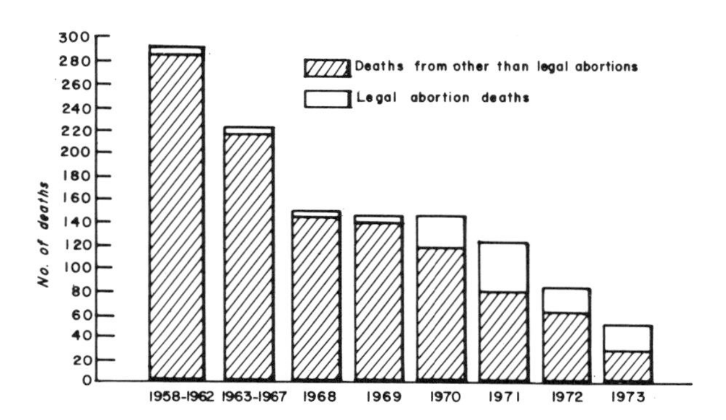

FIGURE 8-7. Deaths associated with abortion, U.S., 1958 to 1973, by legal status of abortion. (Adapted from Tietze, C.: *The effect of legalization of abortion on population growth and public health*. Fam. Plann. Perspect. 7:123, 1975.)

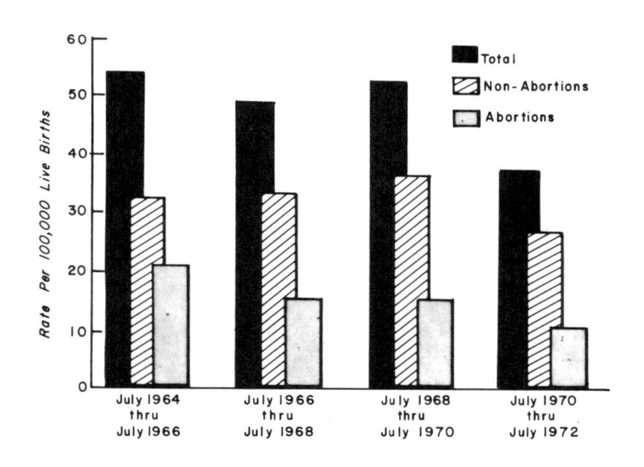

FIGURE 8-8. Puerperal mortality by two-year period, July 1961 to June 1972. (Adapted from Pakter, J., et al: *A review of two years' experience in New York City with the liberalized abortion law*. In Osofsky, H. J., and Osofsky, J. D. (eds.): *The Abortion Experience*. Harper and Row Publishers, Inc., New York, 1973, p. 64.)

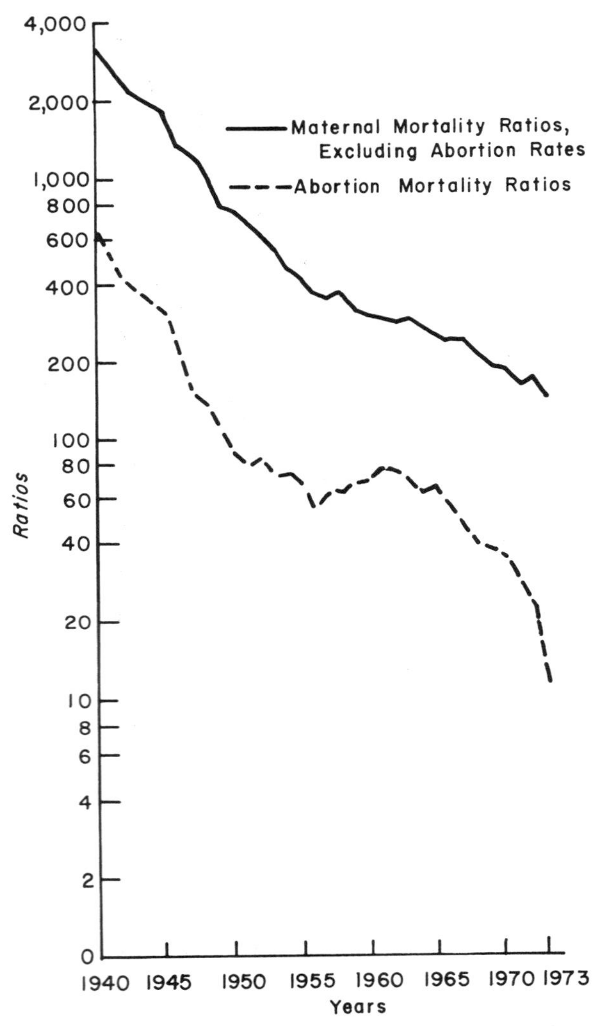

FIGURE 8-9. Maternal mortality ratios (excluding abortion deaths) and abortion mortality ratios, U. S., 1940 to 1973. (Adapted from Oates, W.: U.S. Dept. of Health, Center for Disease Control, 13th Annual Meeting of the Association of Planned Parenthood Physicians, Los Angeles, CA., April 18, 1975.)

In Romania, restrictive abortion laws passed in 1966 reversed previously liberal laws. This resulted in (1) a fivefold increase in abortion related deaths from 1966 to 1970; (2) an increase in the birth rate in one year from 14.3 per 1,000 population in 1966 to 40 per 1000 in 1967; and (3) an increase in the perinatal mortality rate.[7]

Decline in Number of Septic and Incomplete Abortions

Following liberalization of the abortion laws, a universal decrease in the rates of septic and incomplete abortions has been noted. Many of

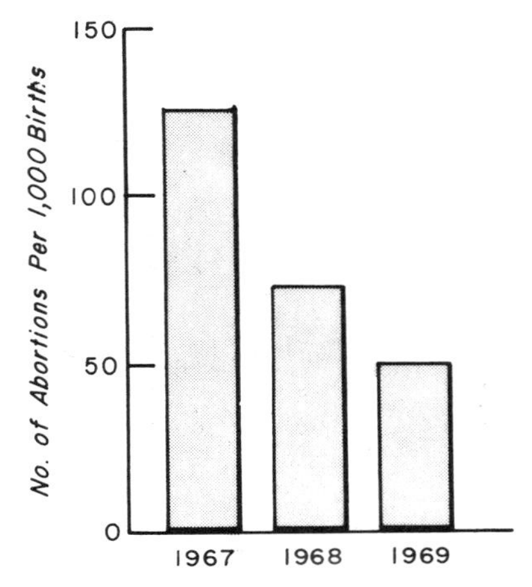

FIGURE 8-10. Nonseptic abortion rate in a private hospital. (Adapted from Steurt, G. L., and Goldstein, P. J.: *Therapeutic abortion in California. Effects on septic abortion and maternal mortality.* Obstet. Gynec. 37:510, 1971.)

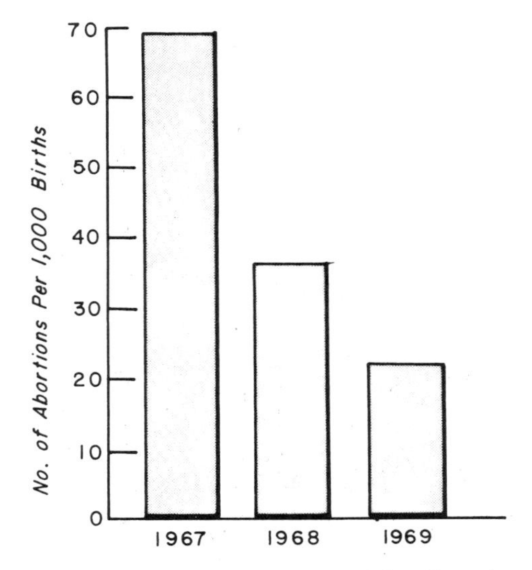

FIGURE 8-11. Septic abortion rate of San Francisco General Hospital. (Adapted from Steurt, G. L., and Goldstein, P. J.: *Therapeutic abortion in California. Effects on septic abortion and maternal mortality.* Obstet. Gynec. 37:510, 1971.)

Table 8-4. Impact on ancillary health care services.

	1969	*1971–1972*
No. of spontaneous abortions	264.0	153.0
Percent of deliveries	16.3	7.9
No. of induced (septic) abortions	91.0	6.0
Postpartum psychoses	0.25	0.10
No. of surgical sterilizations	44.0	109.0
Percent of deliveries	2.8	5.6
Percent of Family Planning Clinic, new patients	1,371	2,960

Adapted from Rovinsky, J. J.: *Impact of a permissive abortion statute on community health care.* Obstet. Gynecol. 41:781, 1973.

the previous septic and incomplete abortions were the results of clandestine procedures (Figs. 8-10 and 8-11). Table 8-4 illustrates the impact of legalized abortion on ancillary health care services.

Prior to legalization of abortion, the cost of caring for 5700 females admitted to New York City hospitals in 1965 following criminal abortions was approximately three and a half million dollars. This could have financed early abortion for over 20,000 females.[8]

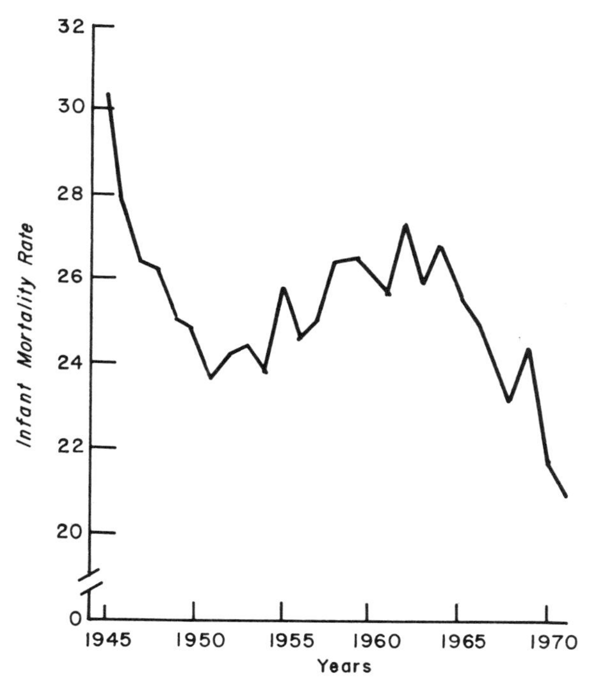

FIGURE 8-12. Infant mortality rates in New York City, 1945 to 1971. (Adapted from Pakter, J., et al: *A review of two years' experience in New York City with the liberalized abortion law.* In Osofsky, H. J., and Osofsky, J. D. (eds.): *The Abortion Experience.* Harper and Row Publishers, Inc., New York, 1973, p. 67.)

Decrease in Infant Mortality (Fig. 8-12)

The obvious decrease in infant mortality probably is related to the reduction in the number of illegitimate births, which are frequently associated with higher mortality rates. (There has been a 16 percent reduction in illegitimate births in New York City and 12 percent in California following abortion legalization[9]).

As a corollary to this there has been a decline in the number of deliveries without prenatal supervision and subsequently a decline in the number of low birth weight infants.

Decline in Child Abandonment

In New York City there was a 56 percent decline in abandoned or adopted infants in one large municipal hospital following abortion liberalization, and there was a 41 percent decline in the number of children being placed for foster care.[10]

Decrease in Out-of-Wedlock Births

Approximately 75 percent of women obtaining abortions are single. There was a continued rise in the number of out-of-wedlock births in New York City until passage of a nonrestrictive abortion law in 1971.[11] There may also be a related decrease in the number of births that were legitimized by marriage.[12]

Table 8-5. Estimate of first year additional government costs for health and social welfare if Medicaid reimbursement for abortion were not available, by type of cost, New York City.

Type of cost	*Unit cost*	*No. requiring care*	*Total cost (in 000s of dollars)*	
			At 10,000 births	*At 15,000 births*
Septic/incomplete abortions	600	5,000	$3,000	
Medical care, full-term pregnancy (total)	*	*	$17,470	$26,205
Prenatal care	400	10,000–15,000	4,700	7,050
Intrapartum hospitalization	611	10,000–15,000	8,110	9,165
Normal new born hospitalization	270	8,000–12,000	2,160	3,240
Premature high risk hospitalization	2,250	2,000–3,000	4,500	6,750
Social welfare programs (total)	*	*	27,099	40,658
Foster-institutional	7,300	485–727	3,641	5,314
ADC and Home Relief (old cases)	980	7,850–11,500	7,342	11,014
ADC and Home Relief (new cases)	3,000	1,550–2,350	4,701	7,050
Infant day care	5,000	2,300–3,450	11,515	17,280
Well-baby health services (total)	*	*	3,293	4,942
Child health station visits	127	9,200–13,800	1,165	1,749
Supplemental feeding	231	9,200–14,800	2,128	3,193
TOTAL COSTS	*	*	50,862	74,805

From Robinson, M., Pakter, J., and Sugir, M.: *Medicaid coverage of abortions in New York City: cost and benefits.* Fam. Plann. Perspect. 6:202, 1974, with permission.

*Not applicable

Out-of-wedlock births have always meant associated loss of educational and economic opportunity for the mothers involved often with a concomitant increase in the welfare roles. In addition, these births were of enormous cost to the government. Without medical coverage of abortion, an estimated 10 to 15 thousand births would result each year in New York City with additional costs to the government (Table 8-5). This compares to an expenditure of 7.75 million dollars to providers of abortions for Medicaid-eligible New York City resident women in 1973.

Equal Access for All Socioeconomic Groups

Prior to abortion legalization there was a marked discrepancy in the number of abortions performed on the socially deprived (e.g., whites versus non-whites). This discrimination against poor families resulted in a much higher incidence of septic abortions and incomplete abortions in the latter group. With liberalization of the laws and with Medicaid coverage of abortions, equal access to abortion became possible for all women. Although white women obtained two thirds of all abortions, non-white women had abortion ratios about one third greater than white women.

Access to Primary Health Care

All abortion candidates should receive a complete medical screening. This evaluation may reveal conditions which, otherwise, might escape detection. For instance, obtaining routine gonorrheal cultures in asymptomatic patients in Metropolitan Hospital, New York City, revealed 33 cases of gonorrhea in 2000 patients, which otherwise would not have been detected. Pap smear screening detected 190 abnormal Pap smears in 6691 patients: 63 with mild to moderate dysplasia, 5 with severe dysplasia and 18 patients with carcinoma in situ.[13]

Improved Contraceptive Utilization

Of females securing abortions in 1971 in the state of Washington, 70 percent were using no contraception at the time of conception.[14] However, there is convincing evidence that, because abortion and contraception serve the same purpose, the woman who uses contraception is more likely than others to seek termination of pregnancy and the female who has had an abortion is more likely to use contraception.[15]

Since 1971 in New York City there has been an improvement in contraceptive practices and family planning clinic utilization (see Table 8-4). In Japan contraceptive utilization has nearly tripled since abortion legalization in 1948. This has been associated with a concomitant decline in the abortion rate.[16]

Diagnosis of Birth Defects

Use of amniocentesis and cell culture have enabled the physician to detect a number of congenital disorders and genetic defects early in the second trimester of pregnancy. The possibility of prenatal diagnosis and pregnancy termination has prevented the birth of infants with developmental malformations and genetic defects and has enabled many women to proceed with a pregnancy which they might otherwise have avoided (Table 8-6).

Abortion may also be resorted to in cases where the fetus is exposed to nongenetic causes of teratogenicity, e.g., rubella, drugs, radiation. The long term savings to the involved family and to society in general are immense.

Table 8-6. Milunsky survey of North American experience with ammniocentesis for prenatal genetic studies.

Indications	*No. of cases*	*Affected fetus*	*Ratio of affected fetuses to cases*	*Therapeutic abortions*	*Prenatal diagnosis confirmed*	*Normal births delivered*
Chromosomal disorders	1368	36	0.026	32	31	745
X-linked disorders	115	54	0.47*	40	34	39
Metabolic disorders	180	37	0.21	30	26	109
	1663	127	0.03	102	91	893†

From Milunsky, A.: *The Prenatal Diagnosis of Hereditary Disorders,* 1973, p. 37. Courtesy of Charles C Thomas, Publisher, Springfield IL.

*The proportion of affected fetuses with possible X-linked disorders is much higher than other classes of disorders. This is partly due to the assignment of all male fetuses as "affected" in those X-linked disorders in which the disease itself cannot be detected in fetal cells. The presence of a male fetus thus indicates a 50 percent risk the fetus has the disease. When both parents are carriers, the probability of conceiving a fetus with an autosomal recessive metabolic disorder is one in four, close to the 21 percent rate observed in the table. On the other hand, the risk of a chromosomal disorder is much lower, even in the group of selected pregnancies represented in the table.

†Not all of the 1,663 pregnancies had been completed at the time of publication.

POTENTIAL DISADVANTAGES OF LEGALIZED ABORTION

Increase in Abortions

There has been concern voiced that with legalized abortions, there will now come an increase in the number of abortions. With replacement of illegal abortions by legal abortions there has been a presumed increase in the number of abortions performed annually. This is speculative, however, since no one really knows the number of illegal abortions performed prior to liberalization of the laws. For example, among New York City residents, the abortion ratio increased 53 percent between 1971 and 1974 and the rate per 1000 women aged 15 to 44 increased by 28 percent.[17]

The Sole Method of Fertility Regulation

Will abortion be used as the sole method of fertility regulation? As described previously, where contraception is available, contraception utilization increases with legalization of abortion. Abortion by itself without contraception is an inefficient method of fertility regulation. With increased contraceptive use, abortion becomes more efficient but only as a backup method. For example:

1. There is a total fertility rate of 7.0 live births per female during her lifetime without utilization of contraception or abortion.

2. Without contraception it would require 9.6 abortions per female to reduce this rate to replacement fertility (2.1 children).

3. With 95 percent effective contraception, a total abortion rate of 0.7 per female or 318 abortions per 1000 live births would be required to achieve replacement fertility.[18]

4. When abortion is combined with no contraception, 30 to 43 births are averted for every 100 abortions.[19]

5. In a contracepting population (90 percent effectiveness) 72 to 82 births are averted for every 100 abortions.[20]

Repeat Abortions Will Become Common

Will the ease in obtaining an abortion make repeat abortions a common occurrence? This is a serious question because repeat abortions place an obvious strain on already overburdened facilities.

Studies show in those countries where contraceptive availability is poor (e.g., Hungary) repeat abortion is common (60 percent of patients are repeaters).[21] However, in areas where contraception utilization and effectivenes is high and where post-abortion contraception counselling and services are available, the incidence of repeat aborters is low[22] (e.g., among New York City residents in 1974, 16 percent of abortion patients were repeaters[23]). Nonetheless, a significant incidence of repeat abortion would be expected to occur even among females using moderately effective contraception.[24]

As can be seen in Table 8-7 and Figure 8-13, ovulation after abortion generally occurs prior to menstruation and often within two weeks of the procedure. Moreover, a high percentage of patients desiring contraception who were not given or prescribed a contraceptive modality at the time of abortion will not return for fertility regulation methods or follow-up. Contraceptive services should therefore be available following abortion. It should be noted that intrauterine device insertions following dilatation and evacuation (D & E) suction procedures were not associated with an increased complication rate.[25]

Table 8-7. Reproductive function after abortion (return of ovulation and menstruation).

1. After abortion (including ectopic pregnancies): In about 60% ovulation occurs within 2–3 weeks; a luteal endometrial pattern may be observed as early as 7 days after abortion.
2. Average return of ovulation following abortions occurring between:
   - 8–15 weeks of gestation—within 2–3 weeks after abortion.
   - 16–20 weeks of gestation—within 4–6 weeks after abortion.
3. Average return of menstruation following abortions occurring between:
   - 8–15 weeks of gestation—within 4–5 weeks after abortion.
   - 16–20 weeks of gestation—within 6–7 weeks after abortion.
4. In 75–90%, the first menstruation following abortion is ovulatory and continues to be ovulatory in subsequent cycles. In the remaining cases, menstruation is anovulatory followed by ovulatory menstruation a month later. *Note*: In 5–10% of healthy nonpostpartum women, anovulatory cycles are observed.

Adapted from Vorherr, H.: *Contraception after abortion and post partum*. Am. J. Obstet. Gynecol. 119:1002, 1973.

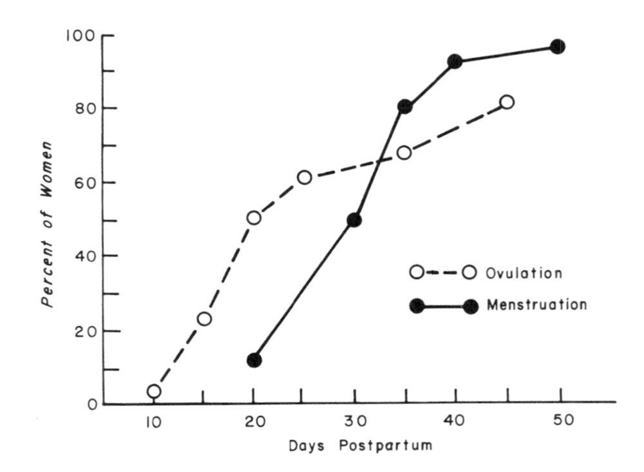

FIGURE 8-13. Return of ovulation and menstruation after abortion. (Adapted from Vorherr, H.; *Contraception after abortion and post partum*. Am. J. Obstet. Gynecol. 119:1002, 1973.)

Psychological Damage to the Patient

In considering the psychological impact of abortion on the patient, one must compare the emotional trauma of the abortion procedure with both the trauma of the events leading up to the abortion and the trauma of undesired motherhood. One must also consider the future of the unwanted child and the psychological impact on him.

In a Swedish survey of children born to mothers denied abortion, it was discovered among these children an increased need for psychiatric help, an increase in antisocial and criminal behavior and an increase in welfare dependency.[26]

The Osofskys in their study of the psychological impact of abortion noted that

> For most women, abortion has had few, if any, psychological sequelae. In the limited number of cases where feelings of guilt or depression have been present, they have tended to be mild and transient in nature. On the whole, the experience has led to further emotional maturation and resolution of conflict. In the rare instances where psychiatric disturbances have been noted postabortion, they have appeared related to existent psychopathology rather than to the procedure.[27]

It appears that more often abortion leads to individual growth and resolution of problems (Tables 8-8 and 8-9). However, it should be noted that there was an increased difficulty in decision-making and more postabortal feelings of guilt and depression among individuals undergoing second trimester abortions.

Table 8-8. Psychological evaluation in 580 cases.*

Variable	*Rating*	*Percent*
Predominant mood		
very unhappy	1	3
moderately unhappy	2	9
neutral	3	18
moderately happy	4	25
very happy	5	45
Physical emotionality		
much crying	1	5
moderate crying	2	5
neutral	3	17
moderate smiling	4	28
much smiling	5	48
Feelings about abortion		
negative: much guilt	1	5
moderate guilt	2	10
neutral	3	12
moderate relief	4	17
positive: much relief	5	56
Attitudes toward self		
negative: angry	1	1
moderate anger	2	5
neutral	3	10
moderate happiness	4	22
positive: happy	5	63

From Osofsky, H. J., and Osofsky, J. D.: *Psychological effects of abortion: with emphasis upon immediate reactions and follow-up.* In Osofsky, H. J., and Osofsky, J. D. (eds.): *Abortion Experience.* Harper & Row Publishers, Inc., New York, 1973, p. 197, with permission.

*Ratings from Negative one to Positive five.

Table 8-9. Follow-up of patients post-abortion.

Patient's report	*Percent 4 wks.*	*Percent 6 mos.*
Reflection on decision		
Satisfaction	94	85
Doesn't know/hasn't thought about it	4	10
Dissatisfaction	2	5
Aftereffects of nonincapacitating nature		
Bleeding and/or cramps	19	24
Psychological	4	5
Physical and psychological	3	4
Neither	74	67
Postoperative interval to resumption of full activities		
Immediately–2 days	78	72
3–7 days	13	14
8–14 days	3	7
15 days–1 month	4	3
Greater than 1 month	2	4

Adapted from Osofsky, H. J., and Osofsky, J. D.: *Psychological effects of abortion: with emphasis upon immediate reactions and follow-up.* In Osofsky, H. J., and Osofsky, J. D. (eds.): *Abortion Experience.* Harper & Row Publishers, Inc., New York, 1973, p. 199.

Threat to Health and Future Reproduction

There have been suggestions that abortion is detrimental to the patient's health and her future reproductive capability. Statistically there is no hard data supporting secondary infertility following abortion. Nevertheless, sterility subsequent to abortion may and does occur as a result of several iatrogenic sequelae. These include Asherman's syndrome, tubal occlusion, infectious complications and uterine perforations with subsequent scars. As for the possibility of increased risk of subsequent premature births and second trimester abortions, conflicting data exists. Caution must be taken in analyzing the results because of the limitations of the method used.[28] The validity of the retrospective studies supporting this concept is questionable since a woman whose pregnancy ended unfavorably would be more likely to admit to a prior induced abortion than a woman with a favorable outcome.

Sensitization of an Rh negative female is an avoidable complication of therapeutic abortion. One should be aware that (1) Detection of fetal erythrocytes in maternal circulation following fetomaternal hemorrhage was noted in 7.2 percent of patients following D & E suction; (2) The incidence of fetomaternal hemorrhage increases with increasing gestation from 1 percent at 8 weeks to over 20 percent at 20 weeks; (3) Amniocentesis itself will cause a fetomaternal hemorrhage in a high percent of patients; and (4) Rho(D)immune human gamma globulin RhoGAM® should be used in Rh negative females and Coombs titers followed.[29]

RISKS AND COMPLICATIONS OF THERAPEUTIC ABORTION

Most of the following data was obtained from the Joint Program for the Study of Abortion (JPSA) in which 72,988 abortions were studied in 66 participating institutions from July 1, 1970, through June 30, 1971,[30] and from the Center for Disease Control, National Abortion Death Registry.[31]

Mortality

In analyzing mortality statistics one should compare the mortality of therapeutic abortion with that of term pregnancy, illegal abortion and with other forms of contraception (Table 8-10).

Table 8-10. Birth control and maternal mortality rates.

Birth control method (1 million users)	*Failure rate*	*Pregnancies*	*Women of all ages Annual death due to:* Pregnancy	Method	Total
Pill	1.0	11,000	3	20	23
IUD	5.0	48,000	12	NK*	NK*
Condom	10.0	106,000	28	—	28
Coitus interruptus	17.0	170,000	44	—	44
Diaphragm	20.0	200,000	52	—	52
Safe period	23.0	231,000	60	—	60
Douche	31.0	310,000	81	—	81
Therapeutic abortions	—	5,000,000	—	125	125
Criminal abortion	—	5,000,000	—	2,500	2,500
Uncontrolled fertility	—	10,000,000	2,500	—	2,600

From Peel, J., and Potts, M: *Textbook of Contraceptive Practice.* Cambridge University Press, New York, 1970, p. 253, with permission.

*NK—not known

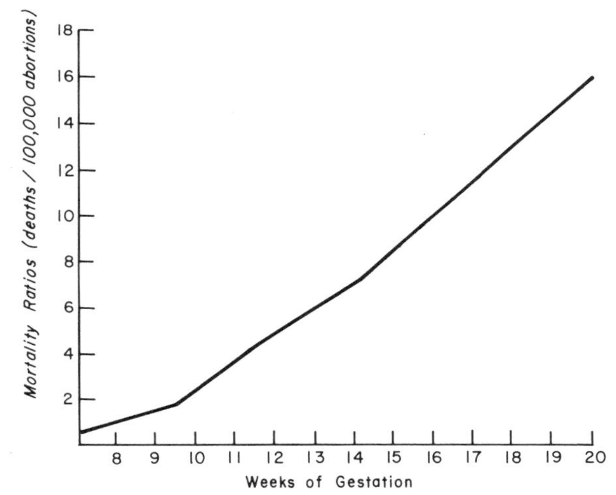

FIGURE 8-14. Abortion mortality ratios by weeks of gestation, U. S., combined 1972-1973 data. (From *Legalized Abortion and the Public Health.* Institute of Medicine, National Academy of Sciences, Washington DC, May 1975, p. 75.)

The mortality statistics vary with the time of the abortion, earlier abortions being much safer than pregnancy (Fig. 8-14) and second trimester abortions having nearly the same mortality. Legal abortion in the first trimester is far less of a hazard to a woman's life than is carrying a pregnancy to term. At the same time, one must consider the hazards of the method used (Figs. 8-15 and 8-16).

Rovinsky's study[32] showed the following facts:

1. mean obstetrical mortality rate of 23 per 100,000 deliveries
2. first trimester abortion mortality of 1.2 per 100,000 abortions
3. second trimester abortion mortality of 17.7 per 100,000 abortions

In over 1.3 million abortions surveyed in the United States in a one-year period (1972 to 1973)[33] there were 45 fatalities with a rate of 3.4 fatalities per 100,000 abortions. The following was also observed[34]:

1. There was a sevenfold increase in mortality after the twelfth week.
2. There was a seven- to ninefold increase in mortality with saline abortions over D & E suction and a hundredfold increase in mortality with hysterotomy.
3. Infection following saline abortions was the most common cause of death.
4. Uterine perforation during a vaginal approach in five-sixth of the cases occurred in gestations beyond week twelve.

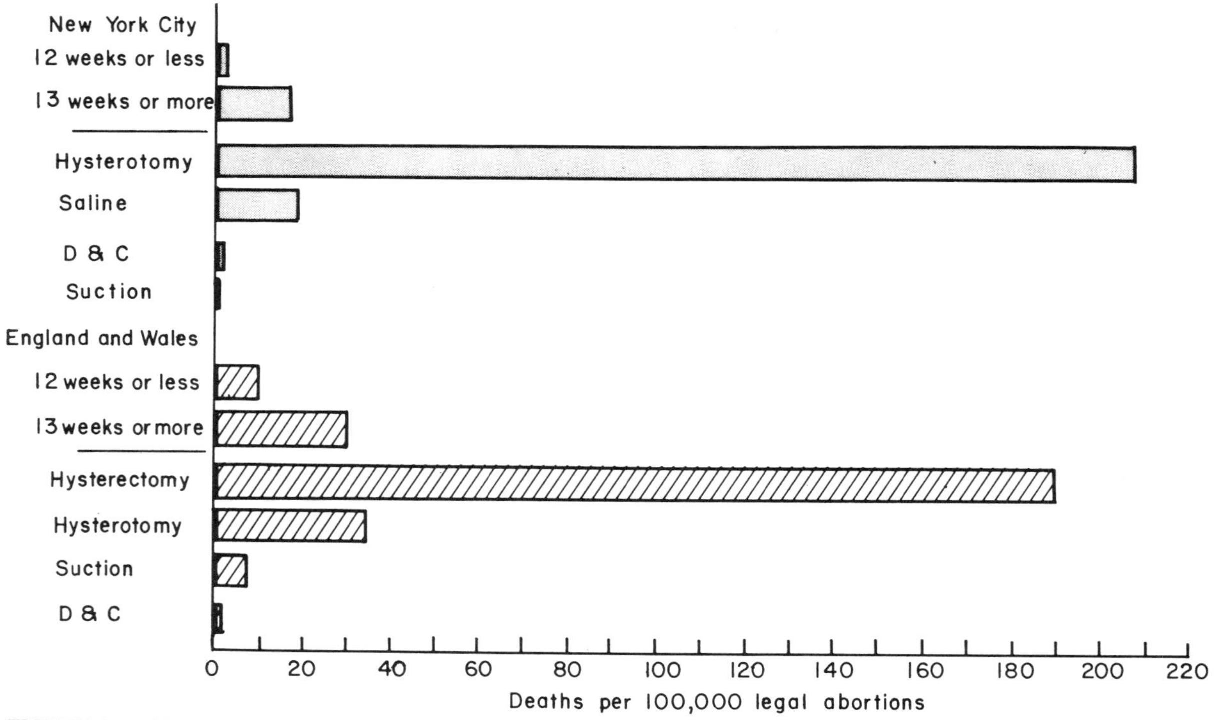

FIGURE 8-15. Mortality per 100,000 legal abortions by gestation and by procedures: New York City and England and Wales, 1972-1973. (Reprinted by permission of The Population Council from Tietze, C., and Dawson, D. A.: *Induced abortion: a factbook*. Reports on Population Family Planning 14:45, 1973.)

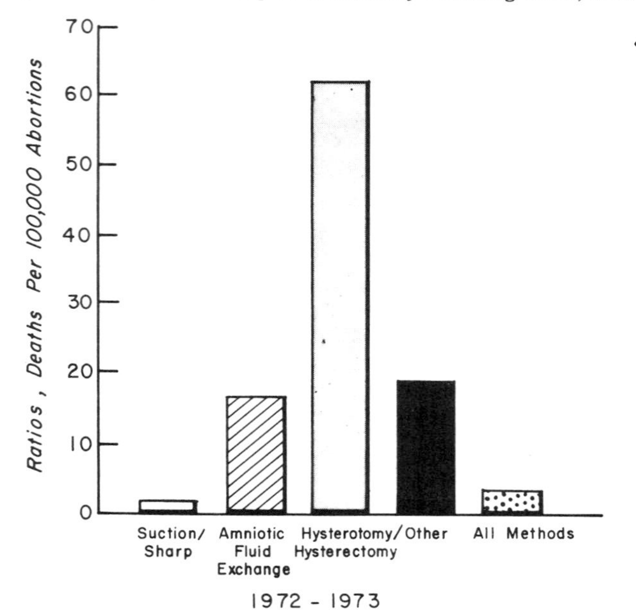

FIGURE 8-16. Death-to-case ratio for legal abortions by method, U. S., 1973 to 1974. (Based on data from *Morbidity and Mortality,* Center for Disease Control, Atlanta, Vol. 24, No. 3, 1975, p. 27.)

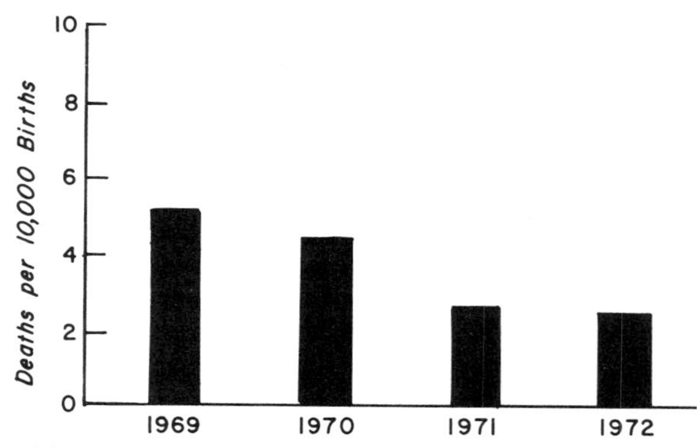

FIGURE 8-17. Abortion related maternal mortality per 10,000 births in New York City (Based on data from Pakter, J., et al.: *A review of two years' experience in New York City with the liberalized abortion law*. In Osofsky, H. J., and Osofsky, J. D. (eds.): *The Abortion Experience*. Harper and Row Publishers, Inc., New York, 1973, p. 47.)

Table 8-11. Mortality ratios of selected surgical procedures, United States, 1969.

Operation	*Ratios*
Legal abortion	
First trimester	1.7
Second trimester	12.2
Tonsillectomy without adenoidectomy	3
Tonsillectomy with adenoidectomy	5
Ligation and division of fallopian tubes	5
Partial mastectomy	74
Cesarean section (low cervical)	111
Abdominal hysterectomy (not abortion)	204
Appendectomy	352

From *Legalized Abortion and the Public Health.* Institute of Medicine, National Academy of Sciences, Washington DC, May 1975, p. 80.

There is a noted drop in abortion related mortality with time in countries with long term abortion practices (Fig. 8-17).

Legal early first trimester abortion carries a lower mortality than most common surgical procedures (Table 8-11).

Morbidity

Again, as we discussed in the section on mortality, the timing of abortion is important. The earlier in gestation the abortion is performed the lower is the subsequent mortality (see Fig. 8-14) and morbidity (Fig. 8-18).

Unfortunately most available pregnancy tests will not diagnose a pregnancy prior to two weeks after the missed menstrual period. Newer methods (radioimmunoassay of the beta sub-unit human chorionic gonadotrophin (HCG) and radioreceptorassay of HCG) make pregnancy detection possible prior to the missed menses.

As time passes, more patients seeking abortion will do so early (Fig. 8-19). The proportion of patients seeking abortion early in pregnancy has been shown to increase with the liberalization of the law.

Delay in seeking an abortion is generally associated with women who are of[35]

younger age
low or very high parity
lower socioeconomic group
unmarried
non-utilizers of contraceptives
incompleted high school education

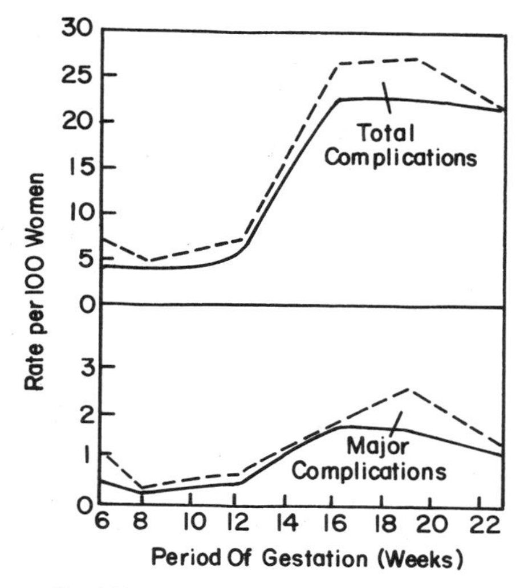

FIGURE 8-18. Range of complication rates per 100 patients with "abortion only," from period of gestation. Note the discrepancies between total and major complications and the importance of close follow-up. (Reprinted by permission of The Population Council from Tietze, C., and Lewit, S.: *Early medical complications of legal abortion.* Studies in Family Planning Joint Program for the Study of Abortion (JPSA) 3:97, 1972.)

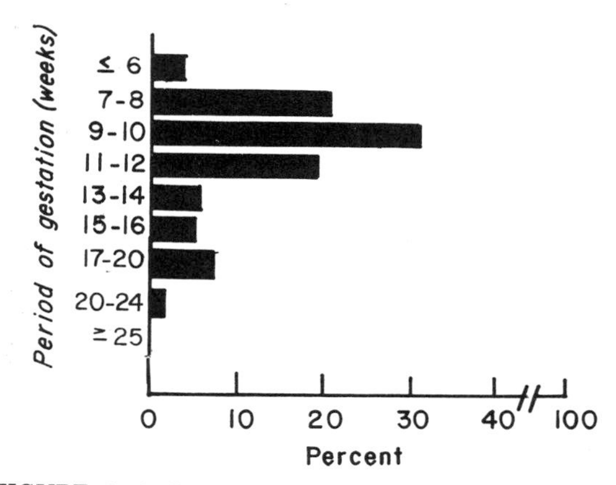

FIGURE 8-19. Percent distribution of patients with induced abortions by period of gestation. (Reprinted by permission of The Population Council from Tietze, C., and Lewit, J.: *Early medical complications of legal abortion.* Studies in Family Planning Joint Program for the Study of Abortion (JPSA) 3:97, 1972.)

These conditions pertain to a group of women with inadequate knowledge, poor contraceptive facilities and an unwillingness to accept the reality of their situations.

Note that those procedures utilized for early pregnancy termination which are most frequently associated with lower morbidity, i.e., suction, are being used with increasing frequency. At the same time there is a decrease in procedures usually used for later pregnancy termination, i.e., saline installation (Fig. 8-20).

Study of Table 8-12 further confirms the lowered morbidity with those procedures associated with early abortions and in the absence of concomitant sterilization. Figure 8-21 shows how complications tend to rise in later

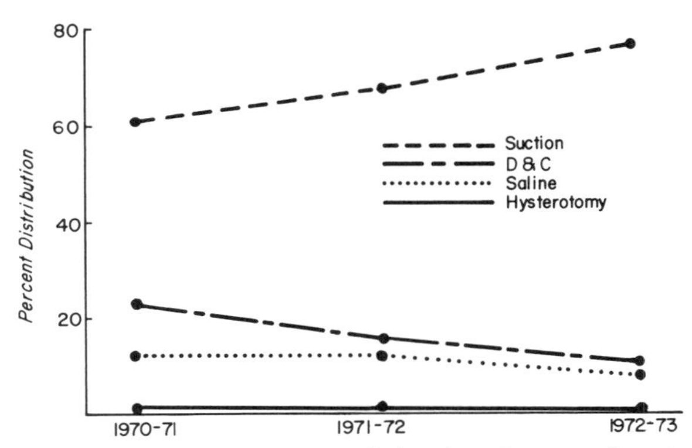

FIGURE 8-20. Distribution of abortions by procedure in New York City. (Based on data from Rovinsky, J. J.: *Impact of a permissive abortion statute on community health care.* Obstet. Gynecol. 41:78, 1973.)

Table 8-12. Number of complications by type of complication and by primary procedure.

Complication	*Suction*	*D & C*	*Saline*	*Hysterotomy*	*Hysterectomy*	*Other*	*Total*
Perforation, uterus	101	22	4	0	0	0	167
including perforation suspected	21	6	1	0	0	0	28
Injury, cervix	495	26	12	0	0	1	534
Injury, intestines	11	0	0	0	0	0	11
Hemorrhage, operational	183	17	0	8	17	0	225
Hemorrhage, postabortal	247	34	0	7	4	1	293
Hemorrhage, unspecified	5	0	373	1	0	4	393
Abruptio placentae	0	0	3	0	0	0	3
Atony, uterine	0	0	2	0	0	0	2
Retention, placenta/tissue	314	27	2,350	3	1	29	2,730
Anemia	84	3	118	20	22	2	249
Endometritis	363	24	389	36	4	10	826
Salpingitis, etc. (P.I.D.)	189	22	53	6	5	0	225
Peritonitis	11	1	5	2	4	0	25
Septicemia	3	0	18	2	1	0	24
Thrombophlebitis	11	0	11	3	10	0	35
Embolism, pulmonary	11	1	0	2	2	0	16
Fever	781	45	1,189	295	358	32	2,700
Anesthesia, complications	80	21	2	7	22	0	132
Convulsions	7	0	7	0	0	0	14
Infection, upper respiratory	9	1	4	1	1	0	16
Infection, urinary tract	109	0	84	44	103	3	349
Obstruction, bowel	1	0	5	12	11	0	29
Pain, cramps	82	30	31	1	1	0	145
Pneumonia	3	0	3	8	11	0	25
Reaction, depressive	39	4	17	3	3	0	60
Reaction, other neurlogic/psychologic	22	1	9	2	3	1	38
Reaction, drug	9	0	2	1	1	0	13
Reaction, saline	0	0	66	0	0	0	66
Shock, syncope, hypotension	16	5	25	1	1	0	48
Wound, disruptive or infection	14	0	1	31	71	0	117

Reprinted by permission of The Population Council from Tietze, C., and Lewit, S.: *Early medical complications of legal abortion.* Studies in Family Planning Joint Program for the Study of Abortion (JPSA) 3:97, 1972.

abortions with or without subsequent sterilization. The major complication rates for patients with or without follow-up are indicated according to type of procedure in Figure 8-22.

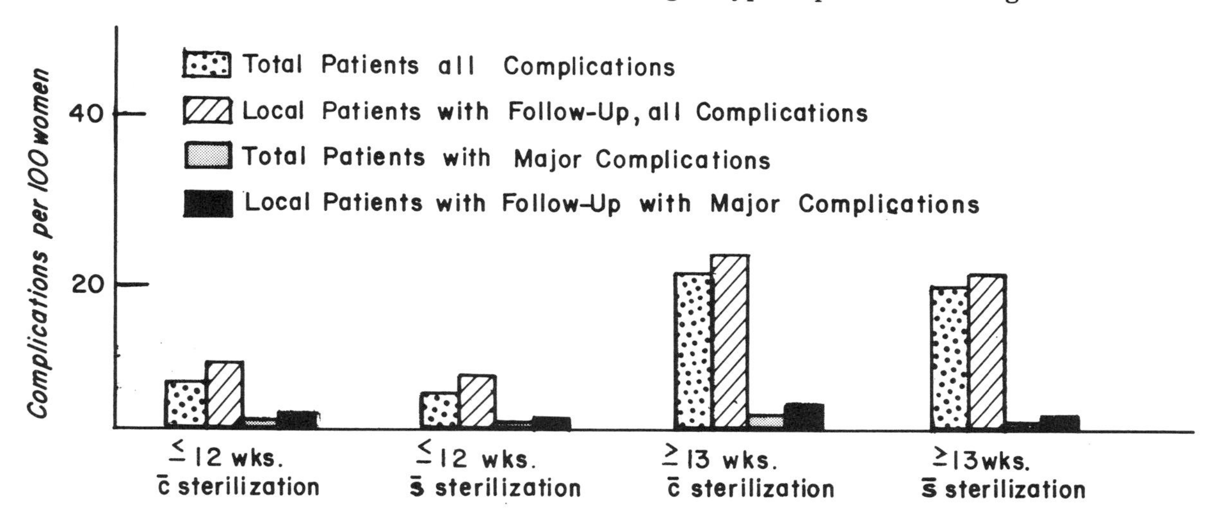

FIGURE 8-21. Abortion total and major complication rate with and without sterilization for all patients and for those with follow-up, as to period of gestation.

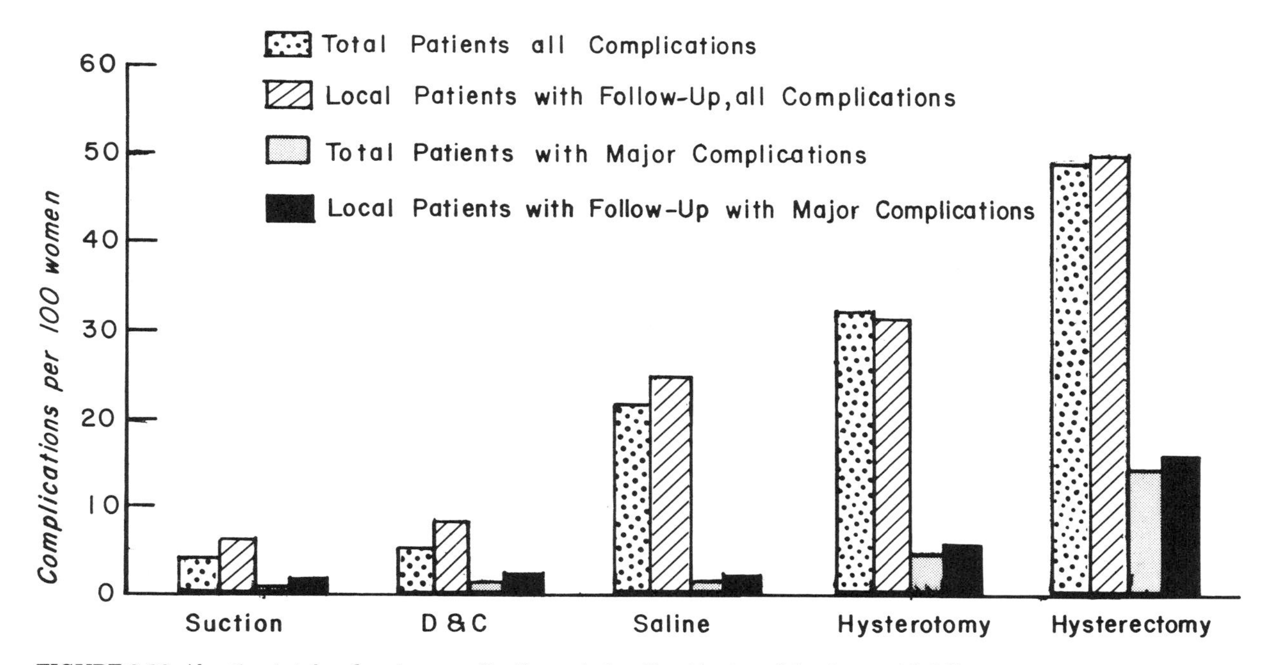

FIGURE 8-22. Abortion total and major complication rate for all patients and for those with follow-up as to type of procedure. (Based on data from Tietze, C., and Dawson, D. A.: *Induced abortion: a factbook*. Reports on Population/Family Planning 14:40, 1973.)

Complications of First Trimester Abortion

In general, complication rates of abortions done in the first trimester are quite low (see Figs. 8-21 and 8-22). Complications are lowest at 7 to 8 weeks gestation and increase steadily thereafter.

The most commonly encountered complications of the first trimester abortion include

- retained tissue
- uterine perforation
- cervical laceration
- pelvic infection
- need for subsequent laparotomy
- continuing pregnancy
- hemorrhage
- transfusion and subsequent complications
- hydatidiform mole
- ectopic pregnancy
- bowel injury

The complication rate increases when abortion is combined with sterilization (see Fig. 8-21) and it is higher when performed under local anesthesia.

Complications of Second Trimester Saline Abortion

As can be seen from Figure 8-22, the incidence of minor complications is quite high (25 percent) while the major complication rate is one tenth of that. Major complications include

- antepartum amnionitis
- endometritis
- postpartum hemorrhage
- need for transfusion
- coagulopathy
- retained placenta
- failed induction
- complications of amniocentesis or saline instillation
- water intoxication and uterine rupture

These will be discussed in more detail later in this chapter. The stage of gestation has no bearing on this complication rate.

Failed Abortion

The failure to successfully terminate a preg-

Table 8-13. Percent distribution of patients with indicated abortions, by primary procedure at first attempt, by outcome at first attempt and by ultimate outcome.

	Outcome of first attempt		*Ultimate outcome abortion*	
Primary Procedure	*Success*	*Failure*	*Same Procedure*	*Different Procedure*
Suction	99.7	0.8	0.1	0.7
D & C	99.3	0.7	0.3	0.4
Saline	95.5	4.5	3.2	1.3
Hysterotomy	100.00	0.0	0.0	0.0
Hysterectomy	100.00	0.0	0.0	0.0

Reprinted by permission of The Population Council from Tietze, C., and Lewit, S.: *Early medical complications of legal abortion.* Studies in Family Planning Joint Program for the Study of Abortion (JPSA) 3:97, 1972.

nancy can and does occur, albeit very infrequently. The failure may be due to the technique used (Table 8-13). In addition to technique failure, one may encounter a failed abortion because of multiple gestations or ectopic pregnancies. The implications for adequate patient follow-up should be obvious.

PATIENT CHARACTERISTICS

It is important to define the type of patient generally requesting an abortion in order to provide better family planning services to that segment of the population. Studies on the typical patient have found her generally to be young, single and a primigravida (Fig. 8-23).

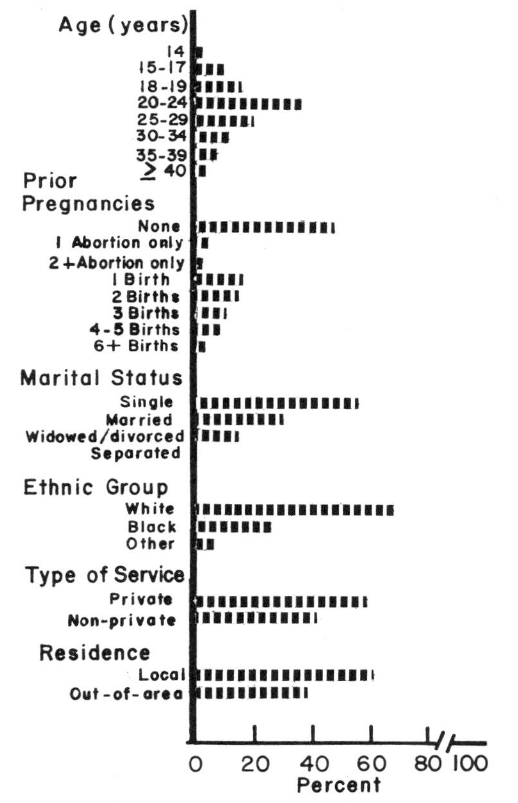

FIGURE 8-23. Percent distribution of patients with induced abortions by age, prior pregnancies, marital status, ethnic group, type of service and residence. (Reprinted by permission of The Population Council from Tietze, C., and Lewit, S.: *Early medical complications of legal abortion.* Studies in Family Planning Joint Program for the Study of Abortion (JPSA) 3:97, 1972.)

1. The age specific ratios of abortions to live births is highest for women less than 20 and those over 35. The difference in the abortion ratio is striking (Figs. 8-24 and 8-25). Related to the high abortion rate under 20 is the fact that nearly 50 percent of illegitimate births are to women under 20.

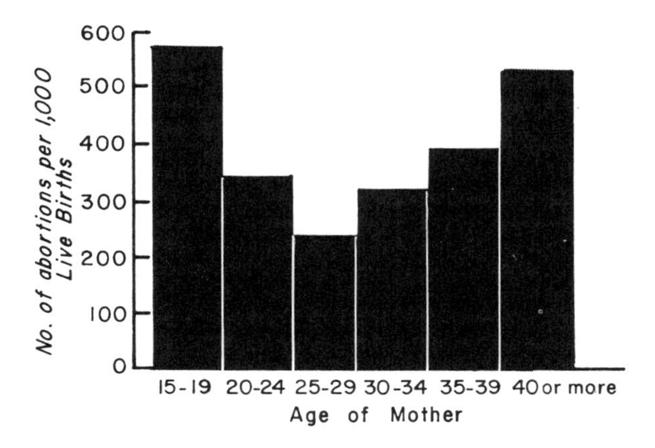

FIGURE 8-24. Abortion ratio for various age cohorts. (Based on data from Tyler, C. W.: *Abortion services and abortion seeking behavior in the United States.* In Osofsky, H. J., and Osofsky, J. D. (eds.): *The Abortion Experience.* Harper and Row Publishers, Inc., New York, 1973, p. 33.)

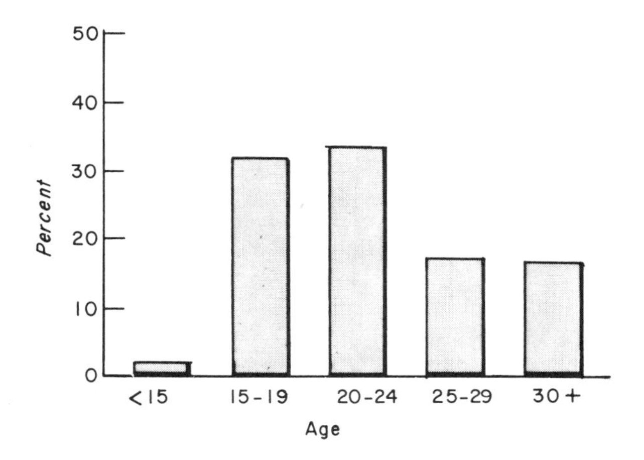

FIGURE 8-25. Percent distribution of reported legal abortions by age, selected states, 1973. (Adapted from *Legalized Abortion and the Public Health.* Institute of Medicine, National Academy of Sciences, Washington DC, May 1975, p. 33.)

2. Abortion in the United States is twice as prevalent among unmarried females (never married—55.9 percent; separated, divorced or widowed—14.2 percent).[36] Looking at the marital status-specific abortion ratios, the abortion ratio for unmarried females was more than 14 times greater than for married females each year[37] (Fig. 8-26).

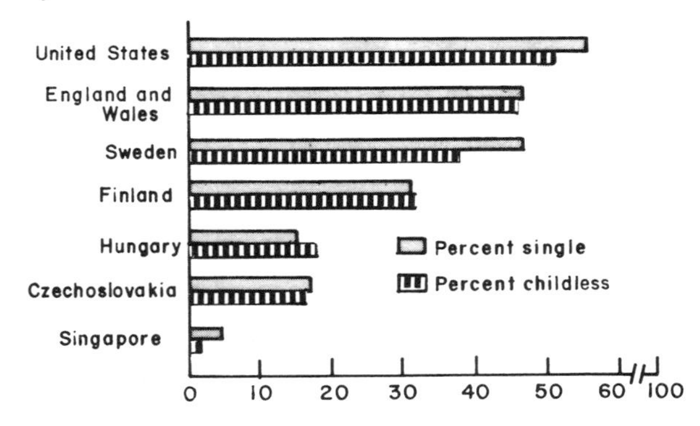

FIGURE 8-26. Profile of women seeking legal abortions, selected countries, 1970 to 1972. (Adapted from Tietze, C., and Dawson, D. A.: *Induced abortion: a factbook*. Reports on Population/Family Planning 14:29, 1973.)

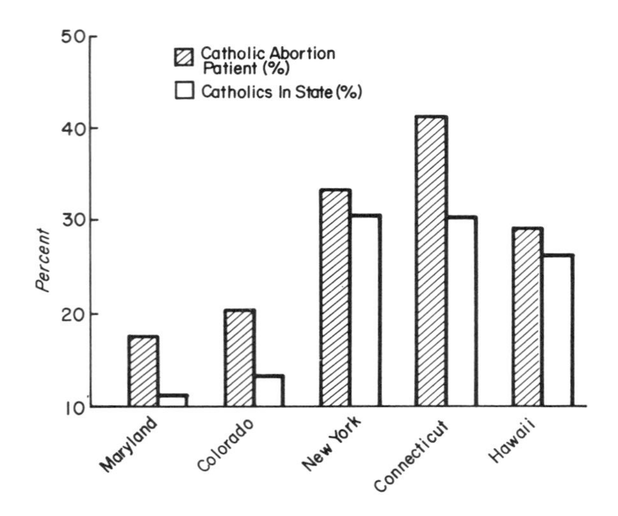

FIGURE 8-27. Percentage of Catholic abortion patients in selected areas compared to percent of Catholics in the state. (Based on data from Sternkoff, P. E.: *Background characteristics of abortion patients*. In Osofsky, H. J., and Osofsky, J. D. (eds.): *The Abortion Experience*. Harper and Row Publishers, Inc., New York, 1973, p. 24.)

3. As can also be seen from Figure 8-26, most abortion patients in the United States are primigravidas.

4. Higher rates of abortion for blacks (as in New York City since liberalization of the abortion laws) probably reflect a difference in successful contraceptive utilization or in the provision of contraceptive services.

5. Despite strong prohibition against abortion by the Roman Catholic Church, the percentage of abortion patients who are Catholic is frequently higher than the percentage of people in the state who are Catholics (Fig. 8-27).

TECHNIQUES OF ABORTION

The type of procedure used for an abortion generally depends upon the stage of gestation. The types of procedure used throughout history are innumerable. The more commonly practiced techniques will be briefly outlined below with emphasis on certain problems that may arise with each procedure.

First Trimester Abortion

D & C

Dilatation and curettage is a simple procedure in which the cervical canal is dilated mechanically and the walls of the uterus are scraped.

Advantages:

1. It can be done as an outpatient procedure.
2. It can be done under local anesthesia (paracervical block) with parenteral analgesia.
3. It requires no sophisticated equipment.

Disadvantages:

1. It is difficult to completely evacuate the products of conception and there may be subsequent hemorrhage and infection.
2. There is risk of uterine perforation.

D & E Suction

Evacuation of the uterine contents with suction was introduced by the Republic of China in

1958 and is currently the most widely used abortion technique. This procedure is done with the patient sterilely prepped and draped in the dorsolithotomy position. The cervix is exposed with a speculum, grasped with a tenaculum, dilated and evacuated under suction pressures of from 30 to 70 mmHg. To better insure complete evacuation, a sharp curettage may be performed and oxytocin or ergotrate may be administered for hemostatis (generally prior to curettage).

Advantages:

1. It is rapid—done as an outpatient procedure.
2. There is a low incidence of complications.
3. It may be done with local anesthesia (paracervical block and parenteral analgesia).

Disadvantages:

1. There is a risk of perforation with the dilator or suction curette with subsequent damage to bowel or mesentery.
2. There is potential damage to cervix with the dilator or tenaculum. This may be obviated with insertion of a laminaria tent 6 to 12 hours preoperatively. Dilation should generally be 1 mm. greater than the number of weeks gestation.

Precautions:

1. Pregnancy tests should be done prior to all first trimester abortions. In 1972 at Metropolitan Hospital in New York City 6.9 percent of patients registered for abortion were found not to be pregnant.[38]
2. Products of conception should be examined grossly. The absence of villi or a fetus may suggest either a perforation or an ectopic pregnancy. Hydatidiform moles can likewise be detected.

Use of Laminaria Tents.[39] Laminaria tents were first reported for use as a cervical dilator in Glasgow in 1862. They are widely used in Japan and currently are increasing in popularity in the United States. They are made from the stems of the brown algae Laminaria digitata. They are gas-sterilized in ethylene oxide and are available in different diameters and lengths. They function by a slow, steady, mechanical dilation of the cervix by hygroscopic swelling (Fig. 8-28).

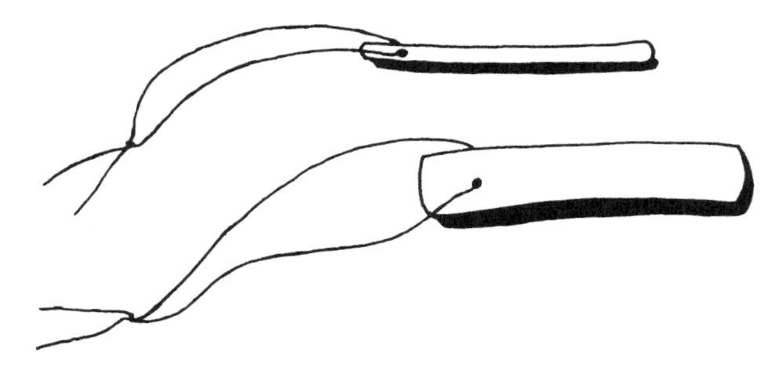

FIGURE 8-28. Hygroscopic swelling of laminaria tent. *Above*, size at insertion; *Below*, swelling after placement.

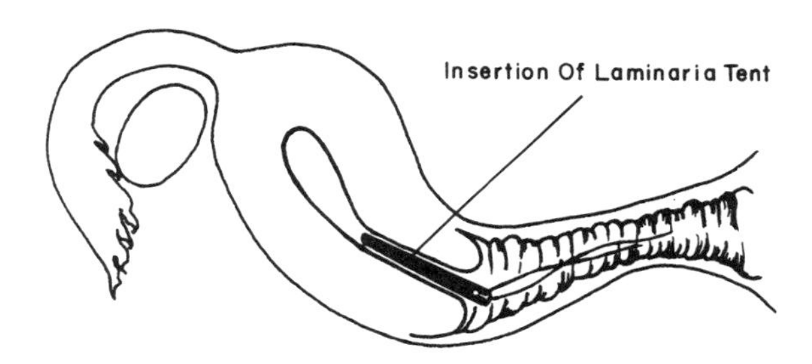

FIGURE 8-29. Placement of a Laminaria tent.

Laminaria tents are generally inserted the day prior to the abortion. They are placed under sterile conditions into the cervical canal just beyond the internal os (Fig. 8-29). (They can be used also in the non-pregnant state for gentle cervical dilation.) A tampon is then inserted and left in place. The laminaria is removed several hours later or the next day just prior to the abortion. Ninety-five percent of patients with a medium-sized laminaria in place for more than 16 hours will show cervical dilation to 13 mm.

Advantages:

1. fewer cervical lacerations
2. fewer uterine perforations
3. shortened operating time
4. less analgesia needed

Disadvantages:

1. occasional expulsion
2. occasional difficulty with removal
3. separate visit to the office required
4. increased infectious complications (adequate statistics lacking)

Menstrual Regulation[40]

Menstrual regulation, endometrial aspiration, mini-abortion or menstrual extraction is performed with a 4 to 6 mm. flexible plastic cannula used under negative pressure (a 20 or 50 cc. syringe can be used in the absence of a suction apparatus) (Fig. 8-30). Cervical dilation or analgesia is not needed. The criteria for performing this procedure is amenorrhea and is generally done within one week of the missed menses.

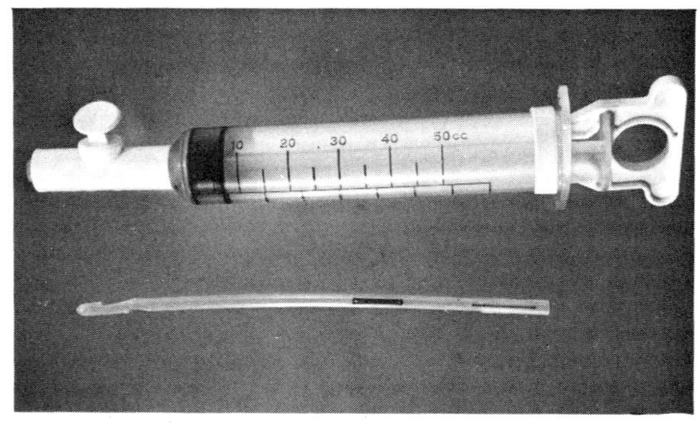

FIGURE 8-30. Equipment for menstrual extraction: syringe with adapter and cannula.

Advantages:

1. rapid
2. easy
3. inexpensive
4. done as an outpatient procedure
5. performed early in gestation with a decrease in anxiety for the patient
6. the patient may resume activities the same day
7. risks, complication rates and failure rates appear to be low (2 to 4 percent) but more extensive evaluation is needed. (See *Disadvantages* for listing of complications.)

Disadvantages:

1. Many patients will be undergoing an unnecessary procedure. In one study pregnancy was documented in only 55 percent of the patients.[41] The percentage could be improved with the use of more sensitive pregnancy tests such as the radioimmunoassay for the beta subunit of HCG or the HCG radioreceptorassay.[42]

2. The effectiveness is not as high as would be preferred—only 95 percent of patients with documented pregnancies had their pregnancies terminated. This was independent of the length of amenorrhea[43,44] (Figs. 8-31 and 8-32).

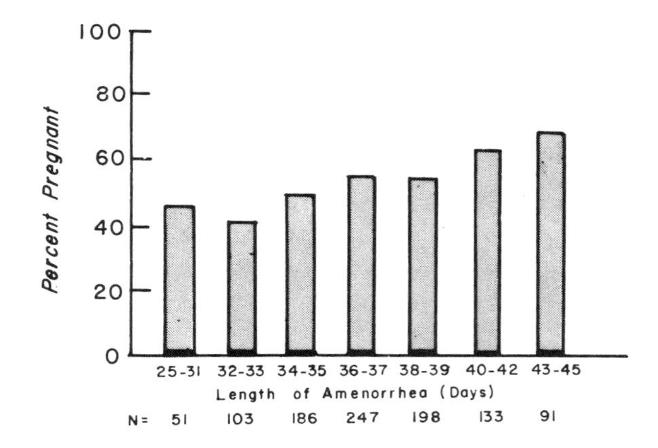

FIGURE 8-31. The proportion in percent of the 1009 women documented to be pregnant (by identification of chorionic villae) at each length of amenorrhea. (Reprinted from Brenner, W. E., et al.: *Menstrual regulation in the United States: a preliminary report.* Fertil. Steril. 26:289, 1975, with permission.)

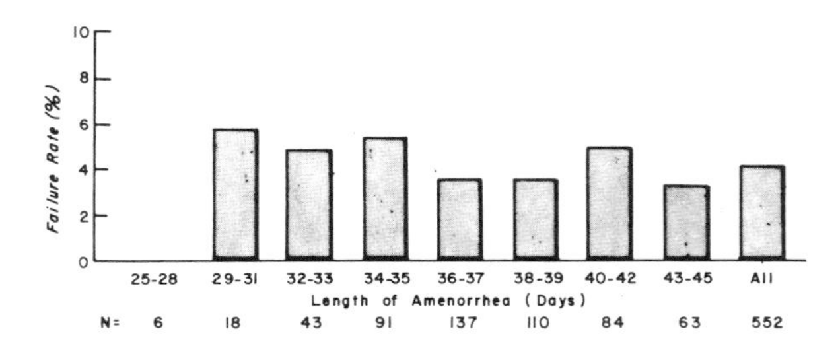

FIGURE 8-32. The failure rate in percent by duration of amenorrhea. (Reprinted from Brenner, W. E., et al.: *Menstrual regulation in the United States: a preliminary report.* Fertil. Steril. 26:289, 1975, with permission.)

3. Complications include perforation, cervical trauma, infection, bleeding, retained products, continuing uterine pregnancies or ectopic pregnancies. Long term complications such as subsequent infertility, cervical incompetence or Rh immunization are yet to be appreciated.

Midtrimester Abortion

Intrauterine Saline Instillation

This has been the most commonly used technique for midtrimester abortion. It involves the instillation of 50 to 250 cc. of a 20 percent sodium chloride solution into the amniotic sac following amniocentesis under local anesthesia. The popularity of this procedure is waning because of the higher proportion of patients seeking earlier abortion and because of the development of better and safer techniques.

Mechanisms of Action. Following the injection there is immediate damage by the hypertonic saline to the decidual cells and chorionic villi and widespread congestion and thrombosis in blood vessels, resulting in subsequent placental and fetal demise. Uterine activity, probably resulting from release of prostaglandins from the decidua, commences within one hour after instillation. Initial contractions are of low amplitude and high frequency. These progress to high-amplitude low-frequency contractions.

Physiologic Changes Following Injection.[45] The following physiologic changes take place:

progesterone—no change
HCG decreases 27 percent
estriol decreases 25 percent
ADH increases
amniotic fluid volume increases
plasma volume—no change or slight increase
hematocrit decreases 10 percent
serum sodium and chloride increase 5 meq./L
serum osmolality increases 15–20 mosml/L
serum potassium—no change
urinary osmolality, sodium and chloride increase and continue elevated for 36 hours
antidiuresis
platelets, fibrinogen, factors 5 and 8, decrease while degredation products increase (secondary to release of tissue thromboplastin into maternal circulation)
prothrombin time—no change
thrombin clotting time prolonged

Disadvantages:

1. It is a difficult procedure prior to week 14 when there is less than 100 cc. of amniotic fluid.
2. The incidence of failure to abort is higher than other methods, being reported as high as 5 percent.
3. The female must carry an unwanted pregnancy more than three months.
4. The injection to abortion interval is long. Unlike the first trimester abortion, saline abortion is slow, with the interval from injection to abortion ranging from 22 to 40 hours (mean 35 hours). (a) The interval is not related to parity or to size of gestation; (b) there is no relationship of interval to amount of amniotic fluid removed; and (c) the interval in 10 to 15 percent of the patients will exceed 72 hours. The use of a concomitant continuous oxytocin drip will decrease the abortion interval to 15 to 34 hours (mean 23 hours) but may increase the complication rate.[46,47,48,49] The use of laminaria tents to shorten the injection to abortion interval has also been suggested. With this, however, the complication rates may also be increased.[50,51,52]

Table 8-14. Complication rate for patients with follow-up after saline abortion.

Complication	*Rate*
Saline specific	0.5
Hemorrhage	15.8
Pelvic infection	2.5
Hemorrhage and infection	3.9
Fever only	3.4
All others	1.1

Adapted from data from Tietze, C., and Lewit, S.: *Early medical complications of legal abortion.* Studies in Family Planning Joint Program for the Study of Abortion (JPSA) 3:97,1972.

5. Complications are high (Table 8-14). For instance,

mortality rate—equal to that of maternal mortality of nonaborted pregnancy (20/100,000)

febrile morbidity (2 to 16 percent of

patients)—subsequent to tissue necrosis and infection

hemorrhage—approximately 2 percent of patients will require transfusion

coagulopathy—noted to a very small extent in every patient, rarely symptomatic, occasionally fatal. Disseminated intravascular coagulopathy has been reported: the treatment for this is to evacuate the uterus and administer heparin.

hypernatremia—secondary to (a) intravascular injection, (b) impairment of renal function, and (c) leakage from uterine puncture site or from amniotic sac into maternal circulation. Symptoms are thirst, burning in mouth, headache, alterations in consciousness, seizure, nausea, flushing, paresthesia of fingertips and hypotension and cardiovascular collapse. Methods of treatment are removal of saline from amniotic sac and intravenous hydration with free water.

myometrial necrosis—secondary to intramural injection of hypertonic saline; patients frequently notice point tenderness in the uterus subsequent to leakage of hypertonic saline into the muscle immediately following injection and removal of the needle; may necessitate hysterectomy.

water intoxication—only if oxytocin is being used and is administered with large quantities of solute poor fluids. The complication results from the antidiuretic action of oxytocin. Symptoms are oliguria, headache, confusion and loss of consciousness. To treat administer hypertonic saline.

retained placenta—occurs in 10 to 15 percent of patients; placenta delivered more than two hours after injection; increased risk of infection and hemorrhage; and routine use of D & C may decrease incidence of complications.

uterine rupture—laceration of posterior cervix; development of cervicovaginal fistula, all related to injudicious use of oxytocin.

failed amniocentesis—must use prostaglandin suppositories or intramuscular analogues as substitutes.

failed onset of labor—greater than 48 hours. Do not reinject supplement with oxytocin or use prostaglandins.

Substitutes

Other substances have been utilized in order to avoid some of the problems associated with hypertonic saline. These include a 50 percent solution of dextrose and urea. The dextrose solution has generally been unacceptable because of increased infection rates, even despite the use of prophylactic antibiotics. Urea has fewer side effects than saline. There is no danger with urea when given intravenously or intraperitoneally. Further studies are needed to evaluate its effectiveness.[53]

Prostaglandins

Prostaglandins are a group of 20 carbon unsaturated hydroxy fatty acids. Fourteen are naturally occurring and found in most, if not all, human tissue. They are readily synthesized from essential fatty acids. Synthetic derivatives are being developed. They function intracellularly and locally and are not found in the systemic circulation. When given intravenously their half life is less than one minute. They function as possible intracellular regulators of metabolic processes through a modification of cyclic-AMP production.

The use of prostaglandins for abortion was first reported in 1970. They are effective at any stage of gestation and cause abortion basically by their ability to initiate myometrial contractions. Prostaglandin $F_2\alpha$ and E_2 have generally been used for this purpose because of their greater specificity for the uterus (Fig. 8-33).

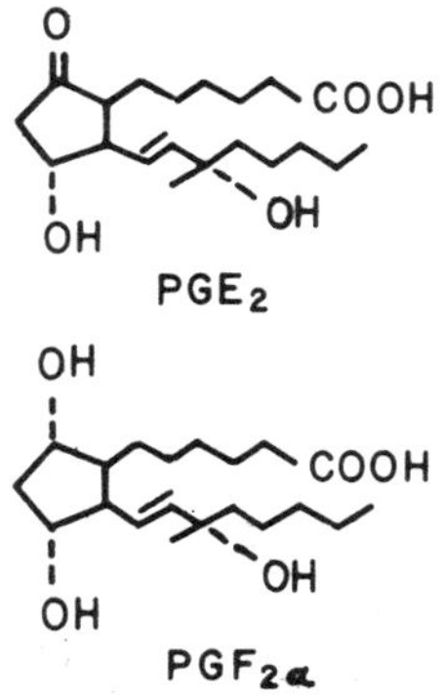

O COOH OH OH

PGE_2

OH COOH OH OH

$PGF_{2\alpha}$

FIGURE 8-33. Prostaglandins E_2 and $F_2\alpha$.

Prostaglandins have offered no advantage over D & E suction for first trimester abortion and therefore their use has been restricted to midtrimester abortions. Routes of administration have included intra-amniotic, intrauterine extra-amniotic, intravaginal, intravenous, oral and intramuscular.

Intra-amniotic. Twenty-five to 50 mgm. of prostaglandin $F_2\alpha$ can be administered during amniocentesis. No amniotic fluid need be withdrawn and frequently an intra-amniotic catheter is left in place for repeated injections. The amniotic fluid acts as a reservoir with very slow transfer of prostaglandin into the maternal circulation, thereby minimizing side effects. The half life in the amniotic fluid is 12 to 13 hours. A test dose should first be given to prevent occasional bronchospasm. Concomitant use of laminaria tents or administration of oxytocin have possible benefit in shortening the injection to abortion interval.

Advantages:

1. Injection to abortion interval generally shorter than with saline (Fig. 8-34)

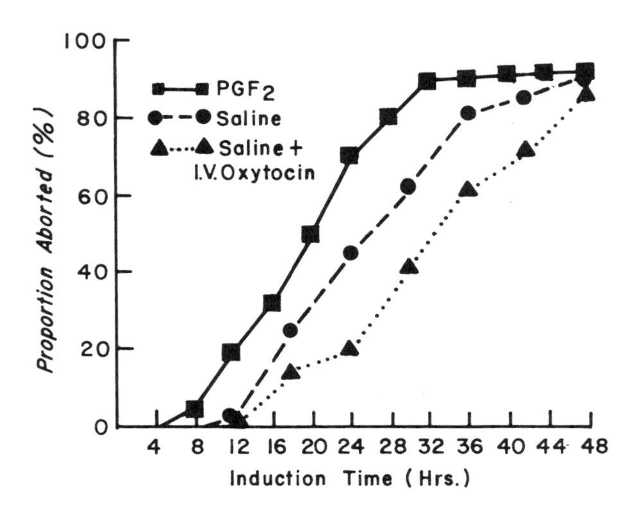

FIGURE 8-34. Cumulative abortion rates of $PGF_2\alpha$ saline, and saline and oxytocin. (Adapted from Brenner, W. E.: *The current status of prostaglandins as abortifacients.* Am. J. Obstet. Gynecol. 23:306, 1975.)

2. Side effects less than with most other means of prostaglandin administration
3. No coagulation defects
4. No risk of hypernatremia
5. No danger of intramyometrial or intravenous injection

Disadvantages:

1. higher incidence of side effects than with saline. Approximately 50 percent of patients will have gastrointestinal symptoms, generally severe, including nausea, vomiting and diarrhea. Other side effects are headache and pyrexia.
2. frequent need for repeat administration (approximately 50 percent)
3. occasional occurrence of lacerations of the posterior cervix
4. other problems associated with midtrimester abortion
5. failure rate slightly greater (8 to 13 percent) than with saline[54,55]

Intrauterine Extra-amniotic. Prostaglandins may be administered through a transcervically-placed polyethylene catheter.

Advantages:

1. lowest incidence of side effects
2. lower dosage required than with intra-amniotic administration

Disadvantages:

1. difficulty in placement of catheter
2. increased risk of sepsis

Intravaginal. Prostaglandins may be administered intravaginally by suppository or pessary.

Advantages:

1. ease of administration
2. potential for administration after missed menses

Disadvantages:

1. high systemic absorption with increased side effects
2. increased failure rate
3. increased incidence of incomplete abortion[56,57]

Intravenous. Intravenous administration of prostaglandins has produced unreliable results and an unacceptable incidence of side effects: diarrhea, nausea and vomiting, headache, pyrexia, flushing, depression and phlebitis.

Oral. The medication is unpalatable and the gastrointestinal side effects are severe. Effectiveness is diminished.

Intramuscular.[58,59] Because the administration of prostaglandin $F_2\alpha$ and E_2 intramuscularly are too painful to be practicable, a synthetic analogue 15(S)-15-methyl-prostaglandin 2-methyl ester has been used for administration by the intramuscular route.

Advantages:

1. ease and safety of administration
2. can be administered with ruptured membranes
3. high success rate

Disadvantages:

1. side effects: vomiting, diarrhea, shivering, pyrexia and anxiety
2. requires repeated injections

The use of prostaglandins for midtrimester abortion is in the current phase of development. With the discovery of synthetic analogues that have greater uterine specificity and lowered side effects and with improved techniques of administration, prostaglandins will play a more dominant role in midtrimester abortions. Nevertheless, the increase complication rate of second trimester abortions, be it saline or prostaglandins, makes these procedures less than ideal, depsite enthusiastic claims.

Hysterotomy

The high mortality (200 deaths per 100,000 legal abortions) and morbidity (33 percent minor, 7 percent major) associated with hysterotomy, relegates this technique to those seldom occasions when other techniques cannot be performed. Subsequent pregnancies must be delivered by cesarean section.

Hysterectomy

A 50 percent minor and 15 percent major complication rate associated with hysterectomy for abortion also makes this generally an unacceptable technique. Its use should be reserved for those instances in which the patient requests sterilization and there are significant pelvic pathological factors to justify it.

Rivanol

Rivanol is a disinfectant used in Japan for midtrimester abortion. Thirty to 50 cc. of a 0.1 percent solution is injected through a catheter into the extraovular space. Oxytocin is given concomitantly. The failure rate is 4 percent and the overall morbidity is low.

Bougie

A bougie is an elastic gum 5 to 10 mm. in diameter and 40 cm. long. It is used in Japan for midtrimester abortion. After cervical dilation to 8 mm. the bougie is inserted into the extraovular space. The extrauterine portion is cut off at the cervical os and the stump is wrapped in a disinfectant gauze; with the onset of uterine contractions the bougie is removed and oxytocin is administered. Failure rate is high (25 percent) and the fetuses are born alive, thereby making this generally unacceptable.

Metreurynter

Another midtrimester abortion technique used in Japan is the metreurynter. A rubber tube with a balloon on the end is introduced through a cervix dilated to 12 to 16 mm. The balloon is filled with 100 to 300 cc. of saline and suspended with weights of 300 to 800 gm. over the foot of the bed. Antiobiotics are administered prophylactically. There is a 15 percent failure rate and the fetuses are born alive.

CONCLUSION

Abortion requires the interplay of the recipient, the community and the provider of the service. The *patient* is confronted with a difficult emotional situation, a situation that requires sympathy, understanding and good counselling. She should be presented with the relative risks and the benefits of the abortion and should not be coerced or persuaded either way. The decision should be hers alone.

The *community* has the responsibility of seeing that the services are provided, meet the highest standards of safety, and are readily accessible to all members of the community.

As for the *provider* of those services, the patient's trust in that institution should not be violated. Abortion services should be integrated into the general obstetric, gynecologic or surgical program of every community hospital. There should be no distinction between abortion and other health care services. The physician should appreciate the importance of the service he or she is providing to that patient and should not develop a negative attitude toward his/her role. Proper counselling should allow the patient to accept the choice of having the child or terminating the pregnancy early.

The deliverer of the health care should be nonjudgmental in the care of the individual requesting the pregnancy termination. He/she should be guided by the conscience of the individual, even when at variance with his own. Under these circumstances, the deliverer of health care should withdraw from active participation but must not abrogate the responsibility for the referral of the woman to local agencies who can better resolve the patient's problem.

Finally, one must appreciate that abortion is a solution for societies' failures. For this reason, the patient should not be held responsible. The best way to reduce the abortion rate is to eliminate those factors that make abortion necessary. This should include enlightened social and sex education in the schools, improvement of communication within the homes, an increase in the number and accessibility of family planning services, and finally improved maternal, infant and child care. Until the development of more universally acceptable contraceptives, however unlikely, together with marked changes in social attitudes and the improbability of total elimination of human failure and irresponsibility, the need for abortion services will continue.

SUMMARY

Abortion is one of the most commonly practiced surgical procedures second only to tonsillectomy. The medical and social advantages of legalized abortion have become evident since the liberalization of the abortion laws in this country. The attitude of most nations has been shifting gradually in favor of supporting legal abortions; however, there has been a concomitant increased resistance by anti-abortion forces. This chapter discusses fairly the advantages and disadvantages as well as the techniques and complications of abortion. Abortion is a solution for society's failures; therefore the patient should not be held personally responsible. To be effective, abortion requires the interplay of the recipient, the provider of the service and the community.

QUESTIONS

1. What effect does legalized abortion have on the abortion rate?
2. What are the medical advantages of legalized abortion? The social advantages?
3. How can the abortion complication rate be reduced?
4. What are the potential long term complication rates of abortion?

REFERENCES

1. Tietze, C., and Murstein, M. C.: *Induced abortion: 1975 factbook.* Reports on Population/Family Planning 14:6, 1975.
2. Donayre, J.: *Country profiles.* Population Council, October 7, 1973.
3. Tietze, C., and Murstein, M. C.: op. cit., p. 7.
4. Tietze, C., and Murstein, M. C.: op. cit., p. 13.
5. Weinstock, E.: *Legal abortions in the United States since the 1973 Supreme Court decisions.* Fam. Plann. Perspect. 7:26, 1975.
6. Robinson, M., et al.: *Medicaid coverage of abortions in New York City: costs and benefits.* Fam. Plann. Perspect. 6:202, 1974.
7. Wright, N. H.: *Restricting legal abortion: some maternal child health effects in Romania.* Am. J. Obstet. Gynecol. 121:246, 1975.
8. Hall, R. E.: *Induced abortion in New York City.* Am. J. Obstet. Gynecol. 110:601, 1971.
9. Pakter, J., et al.: *A review of two years' experience in New York City with the liberalized abortion law.* In Osofsky, H. J., and Osofsky, J. D. (eds.): *The Abortion Experience.* Harper and Row Publishers, Inc., New York, 1973, Chapter 3.
10. *Legal Abortion in the United States: Facts and Highlights.* Planned Parenthood Federation of America, Inc., New York, No. 1442, 1973, p. 10.

11. Pakter, ibid.
12. Sklar, J., and Berkov, B.: *Abortion, illegitimacy, and the American birth rate.* Science 185:909, 1974.
13. Gordon, M.: *First trimester abortion.* In Osofsky, H. J., and Osofsky, J. D. (eds.): *The Abortion Experience.* Harper and Row Publishers, Inc., New York, 1973, p. 369.
14. Fujita, B. W., and Wagner, N. N.: *Referendum 20—abortion reform in Washington state.* In Osofsky, H. J., and Osofsky, J. D. (eds.): *The Abortion Experience.* Harper and Row Publishers, Inc., New York, 1973, p. 254.
15. Potts, P. M.: *Patterns of abortion and contraceptive usage.* R. Soc. Health J. 91:294, 1971.
16. Tietze, C., and Dawson, D. A.: *Induced abortion: a factbook.* Reports on Population/Family Planning No. 14, December 1973.
17. Pakter, J., Nelson, F., and Sugir, M.: *Legal abortion: a half-decade of experience.* Fam. Plann. Perspect. 7:256, 1975.
18. Tietze, C., and Murstein, M. C.: op. cit., p. 61.
19. Ibid.
20. Ibid.
21. Tietze, C.: *The "problem" of repeat abortions.* Fam. Plann. Perspect. 6:248, 1974.
22. Daily, E. F., et al.: *Repeat abortions in New York City:1970–1972.* Fam. Plann. Perspect. 5:89, 1973.
23. Pakter, J., Nelson, F., and Sugir, M.: op. cit., p. 255.
24. Tietze, C.: *The problem . . . ,* ibid.
25. Ibid.
26. Forssman, H., and Thuwe, I.: *One hundred and twenty children born after application for therapeutic abortion refused.* Acta Psychiatric Scandinavia 42:71, 1966.
27. Osofsky, H. J., and Osofsky, J. D.: *Psychological effects of abortion: with emphasis upon immediate reactions and follow-up.* In Osofsky, H. J., and Osofsky, J. D. (eds.): *The Abortion Experience.* Harper and Row Publishers, Inc., New York, 1973, p. 188.
28. *Legalized Abortion and the Public Health.* Institute of Medicine, National Academy of Science, Washington DC, May 1975.
29. Queenan, J. T., et al.: *Role of induced abortion in rhesus immunization.* Lancet 1:815, 1971.
30. Tietze, C., and Lewit, S.: *Early medical complications of legal abortion.* Studies in Family Planning Joint Program for the Study of Abortion (JPSA) 3:97, 1972.
31. *Abortion-related mortality, 1972 and 1973 - United States.* Center for Disease Control, U. S. Public Health Service, Atlanta GA, Vol. 24, No. 3, 1975.
32. Rovinsky, J. J.: *Impact of a permissive abortion statute on community health care.* Obstet. Gynecol. 41: 781, 1973.
33. Tietze, C., and Murstein, M. C.: op. cit., p. 56.
34. Berger, G. S., et al.: *Maternal mortality associated with legal abortion in New York state: 7/1/70–6/30/72.* Obstet. Gynecol. 43:315, 1974.
35. Bracken, M. B., and Swigar, M. E.: *Factors associated with delay in seeking induced abortions.* Am. J. Obstet. Gynecol. 113:301, 1972.
36. Tietze, C., and Lewit, S.: ibid.
37. Tyler, C. W.: *Abortion services and abortion-seeking behavior in the United States.* In Osofsky, H. J., and Osofsky, J. D. (eds.): *The Abortion Experience.* Harper and Row Publishers, Inc., New York, 1973.
38. Gordon, M.: op. cit., p. 368.
39. Hale, R. W., et al.: *Laminaria: an underutilized clinical adjunct.* Clin. Obstet. Gynecol. 15:829, 1972.
40. Van der Vlugt, T., and Piotrow, P. T.: *Menstrual extraction.* Population Report, Series F, No. 4, 1974., p. 49.
41. Edelman, D. A., et al.: *The effectiveness and complications of abortion by dilatation and vacuum aspiration versus dilatation and rigid metal curettage.* Am. J. Obstet. Gynecol. 119:473, 1974.
42. Landesman, R., and Saxena, B.: *Radioreceptorassay of human chorionic gonadotropin as an aid in miniabortion.* Fertil. Steril. 24:1022, 1975.
43. Van der Vlugt: ibid.
44. Brenner, W. E., et al.: *Menstrual regulation in the United States: a preliminary report.* Fertil. Steril. 26:289, 1975.
45. Burnett, L. S., et al.: *Techniques of pregnancy termination.* Obstet. Gynecol. Survey 29:6, 1974.
46. Ibid.
47. Hanson, F. W., et al.: *Oxytocin augmentation of saline abortions.* Obstet. Gynecol. 41:608, 1973.
48. McKenzie, J. M., et al.: *Midtrimester abortion: clinical experience with amniocentesis and hypertonic instillation in 400 patients.* Clin. Obstet. Gynecol. 14:107, 1971.
49. Stim, E. M.: *Saline abortion.* Obstet. Gynecol. 40:247, 1972.
50. Hale, R. W.: ibid.
51. Hanson: ibid.
52. Lischke, J. H., and Goodlin, R. C.: *Use of laminaria tents with hypertonic saline amnioinfusion.* Am. J. Obstet. Gynecol. 116:586, 1973.
53. Burnett, L. S., Wentz, A. C., and King, T. M.: *Techniques of pregnancy termination.* Obstet. Gynecol. Survey 29:30, 1974.
54. Karim, S. M. M., and Sharma, S. D.: *Second trimester abortion with single intra-amniotic injection of prostaglandins* E_2 *or* $F_2\alpha$. Lancet 2:47, 1971.
55. Duerhoelter, J. H., and Gant, N. F.: *Complications following prostaglandin* $F_2\alpha$*-induced midtrimester abortion.* Obstet. Gynecol. 46:247, 1975.
56. Karim, S. M. M.: *Contraception: once-a-month vaginal administration of prostaglandins* F_2 *and* $F_2\alpha$ *for fertility control.* Contraception 3:173, 1971.
57. Karim, S. M. M., and Sharma, S. D.: *Therapeutic abortion and induction of labour by the intravaginal administration of prostaglandins* E_2 *and* $F_2\alpha$. J. Obstet. Gynecol. Br. Commonwealth 78:294, 1971.
58. Brenner, W. E., et al.: ibid.
59. Bieniarz, J., Hunter, G., and Scommegna, A.: *Efficacy and acceptability of 15(S)-15-methyl-prostaglandin* E_2*-methyl ester for midtrimester pregnancy termination.* Am. J. Obstet. Gynecol. 120:840, 1974.

BIBLIOGRAPHY

Brenner, W. E.: *The current status of prostaglandins as abortifacients.* Am. J. Obstet. Gynecol. 123:306, 1975.

Burnett, L. S., Wentz, A. C., and King, T. M.: *Techniques of pregnancy termination.* Obstet. Gynecol. Survey 29:6, 1974.

Dytrych, Z., et al.: *Children born to women denied abortion.* Fam. Plann. Perspect. 17:165, 1975.

Jacobs, D., García, C. R., Rickels, K., and Preucel, R. W.: *A prospective study on the psychological effects of therapeutic abortion.* Compr. Psychiatry 15:423, 1974.

Legalized Abortion and the Public Health. Institute of Medicine, National Academy of Sciences, Washington DC, May 1975.

Osofsky, H. J., and Osofsky, J. D.: *The Abortion Experience.* Harper and Row Publishers, Inc., New York, 1973.

Rovinsky, J. J.: *Impact of a permissive abortion statute on community health care.* Obstet. Gynecol. 41:781, 1973.

Tietze, C., and Dawson, D. A.: *Induced abortion: a factbook.* Reports on Population/Family Planning No. 14, December 1973.

Tietze, C., and Lewit, S.: *Early medical complications of legal abortion.* Studies in Family Planning Joint Program for the Study of Abortion (JPSA) 3:97, 1972.

Tietze, C., and Lewit, S.: *Early medical complication of abortion by saline.* Studies in Family Planning Joint Program for the Study of Abortion (JPSA) 4:133, 1973.

Tietze, C., and Murstein, M. C.: *Induced abortion: 1975 factbook.* Reports on Population/Family Planning 14:1, 1975.

Index

of Subject Matter and Illustrations*

ABANDONMENT, child, decline in, with legalized abortion, 131
Abdominal tubal ligation, 108
Abortifacient agent, map of, *118*
Abortion
  age and, *143*
  age distribution in, *48*
  among Catholics, *144*
  as sole method of fertility regulation, 133
  by period of gestation, *139*
  by procedures, *140*
  community in, 150, 151
  complication rate in, *139*
    by period of gestation, *141*
    by type of procedure, *141*
  D & C in, 144
  D & E suction in, 144
  ethnic groups and, *143*
  failed, 142
  failure rate in, by duration of amenorrhea, *146*
  first trimester
    complications of, 142
    techniques for, 144–146
  future reproduction and, 136
  health and, 136
  health care provider in, 151
  incomplete, septic and, decline in, with legalized abortion, 130
  induced, *139*
  intrauterine instillation in, 147–148
  legalized. *See* Legalized abortion.
  legislation pertaining to, 123–127
    in the United States, 124–127
  marital status and, *143*
  maternal mortality in, decline in, 129–130
  menstrual regulation in, 146
  midtrimester, techniques for, 147–150
  mortality and, *129*
    and maternal mortality rates, *130*
    by gestation and procedures, *138*
    by method, *138*
    by weeks of gestation, *137*
  patient characteristics in, 143–144
  patient in, 150, 151
  prior pregnancies and, *143*
  puerperal mortality and, *129*
  psychological problems in, 135
  repeat, potential for, 134
  reproduction and, future, 136
  saline instillation in, 147–148
  septic and incomplete, decline in, with legalized abortion, 130
  techniques of, 144–150
  therapeutic. *See* Therapeutic abortions.
Abortion rate
  nonseptic, *130*
  prostaglandins and, *149*
  saline and, *149*
  septic, *130*
Acne, oral contraceptives and, 94
Adolescent(s)
  coital patterns of, 48–49, *49*
  fertility control among, 49–53
  knowledge of reproduction and conception among, 53
  number of, *48*
  out-of-wedlock births to, *54*
  sexual patterns of, 47–53
  sexually-active, contraceptive methods and, *52*
  sexually-active never-married, contraception and, race and education and, *50*
  sexually-experienced never-married, females, contraceptive method and, *51*
  skeletal maturation in, oral contraceptives and, 93
  social patterns of
    racial differences in, 49
    sexual patterns and, 47–53
Adolescent pregnancy, 54–57
  consequences of, to children, 55
  extent of, 54–55
  facilities for, 57
  medical consequences of, 55
  social consequences of, 55
Adolescent sexuality, 47–57
Adult illiteracy rates, *2*
Age
  abortion ratio and, *143*
  as a factor in fertility, 2
  as a factor in sterilization, 107
  induced abortions and, *143*
  maternal
    children ever born to, *1*
    Down's syndrome and, *36*
    infant mortality and, *56*
    low-birth-weight infants and, *36*, *56*

*Italic numerals are used to indicate pages where illustrations appear.

Age—*Continued*
of adolescents, contraceptive use and, *50*
of wife, sterilization and, *107*
range in, sterilization and, *107*
Age distribution in induced abortions, *48*
Age groups, increase of, anticipated, in U. S., *25*
Age specific fertility rate, definition of, 5
Amenorrhea
Depo-Provera and, 99
postpill, 92
pregnancy and, *146*
Amniocentesis, failed, with intrauterine saline instillation, 148
Anemia
intrauterine device and, 104
oral contraceptives and, 94
Antispermicidal preparations in contraception, 71–73
Assault, sexual. *See* Rape.
Attitude toward contraception, in failure of, 66

BABY boom affecting population of United States, 24, 25
Band-aid surgery for sterilization, 113
Barriers, changing, to woman's equality, 39–43
Basal body temperature in rhythm method, 68, *69*
Belly-button surgery for sterilization, 113
Benefits derived from oral contraceptives, 94
Birth(s)
in United States, to mothers under 18 and over 35, *35*
out-of-wedlock, to teenagers, 54
to mothers under 17 and over 35, *34*
Birth control pill. *See* Oral contraceptives.
Birth defects. *See also* Teratogenicity.
diagnosis of, with legalized abortion, 133
Birth order
fetal death rates and, *35*
infant mortality and, *33*, *36*
Birth rate,
decreasing, as solution to world population growth, 19
fertility control and, *62*
Bleeding
Depo-Provera and, *98*
intrauterine device and, 104
Blood clots. *See* Thromboembolic disease.
Blood pressure, rise in, oral contraceptives and, 89
Bougie as midtrimester abortion technique, 150
Breast(s)
effects of oral contraceptives on, 81
tenderness of, oral contraceptives and, *83*
Breast cancer
mortality from, *90*
oral contraceptives and, 90
Breastfeeding as a contraceptive method, 70
Bucharest Conference, statement of, 7–8

CANCER
mortality from, *90*
oral contraceptives and, 90
Cardiovascular system, oral contraceptives and, 83
Catholic abortion patients, *144*
Central nervous system, oral contraceptives and, 81
Cerebrovascular accidents, oral contraceptives and, 88
Cervical dilation with laminaria tents, 145
Cervical mucus, examination of, in rhythm method, 69
Cervix
cancer of, mortality from, *90*
effects of the pill on, 81
Changing status of woman, 39–45
Characteristics of patients in abortion, 143–144
Chemical occlusive agents for sterilization, 114
Child abandonment, decline in, with legalized abortion, 131
Children
consequences to
in adolescent pregnancy, 55
as affected by short birth intervals, 34
health of, depending on family size, 32–33
mortality rates for, order of pregnancy and, *32*
Children ever born
by age and race, *1*
by occupation of husband and race, *13*
in United States, *26*
Chloasma, oral contraceptives and, 81
Church as barrier to woman's equality, 40
Coagulopathy with intrauterine saline instillation, 148
Cohort fertility, definition of, 5
Coital frequency, *63*
of sexually-active adolescent females, *49*
with contraceptive methods, *77*
with the pill, *77*
Coital patterns of adolescents, 48–49, *49*
Coitus interruptus, 70
Collection of data and evaluation, as element of demography, 3, 4
Community in abortion responsibility, 150, 151
Complications
in intrauterine saline instillation, 147–148
in vasectomy, 115–117
of first pregnancy, *35*
of first trimester abortion, 142
of second trimester saline abortion, 142
pregnancy related, in extremes of reproductive years, 35
Conception
after oral contraception, *92*
knowledge of
by adolescents, 53
by young unmarried females, *53*
Conception rates following Depo-Provera, *99*
Condom as contraceptive method, 73
Continuation rates of IUDs, 102
Contraception
abortion as sole method of, 133
among adolescents, 49–53
education and, *50*
future considerations for, 117
knowledge of, among adolescents, 53
median parity vs. median age and, *64*
methods of. *See* Contraceptive methods.
race and, *50*
sources for, among adolescents, 52–53
utilization of, among adolescents, 49
Contraceptive failure, 64–67
Contraceptive method(s)
among adolescents, 51
efficacy of, 52
factors which influence, 52
patterns of, 51–53
sources for, 52–53
utilization of, 49
antispermicidal preparations as, 71–73
coital frequency and, *77*
coitus interruptus as, 70
condom as, 73
Depo-Provera as, 98
female sterilization as, 108–114
future considerations in, 117
intramuscular injections as, 98
male sterilization as, 114–117
minipills as, 98
morning-after-pill as, 100

Contraceptive method(s)—*Continued*
physicians' reluctance to discuss, 59
pill-a-month as, 99
polymer implants as, 100
postcoital contraception as, 100
postcoital douche as, 70
prescribing of, 67
progesterone as, 106
prolongation of lactation as, 70
race and, *51*
rhythm, 68
sexually-active teenagers and, *52*
sexually-experienced never-married adolescent females and, *51*
sexually-experienced never-married females and, *52*
sources for, among adolescents, 52–53, *52, 53*
sterilization as, 106–117
weighing risks and benefits of, *119*
withdrawal as, 70
Contraceptive users
failure of, *65*
new, with family-planning programs in developing countries, *106*
Contraceptive utilization
by adolescents, 49–53
by never-married adolescent females, *50*
change in, *63*
effectiveness of, 60–67
factors which influence, by adolescents, 49–51
frequency of, by adolescents, 49
improved, with legal abortion, 132
incentives and, *66*
patterns of, among adolescents, 51–53
Contraindications
absolute, to oral contraceptives, 96
to intrauterine device, 104
to oral contraceptives, absolute, 96
Copper IUD, 105
Creams as contraceptive method, 71–73
Crude birth rate
definition of, 4
of developed and developing countries, *11*
Crude death rate
definition of, 4
of developed and developing countries, *11*
Crude rate of natural increase, definition of, 4
Cultural labels as barrier to woman's equality, 40

D & C AS abortion technique, 144
D & E suction as abortion technique, 144
Data collection in demography, 4
Death rates, maternal, *35*
Declining fertility, economic growth and, *18*
Definitions of demographic terms, 4
Demographic characteristics of females, *12*
Demographic factors
in fertility, 1–6
selected categories in, 1–3
Demographic transition, 5
four stages of, *5*
Demography
categories in, 1–3
elements of, 3–5
terms used in, definitions of, 4
theoretical issues in, 3–4
Dependency ratio, definition of, 5
Dependency ratios, high, as world population problem, 15
Depo-Provera, 98
bleeding patterns and, *98*
conception rates and, *99*
Determination of safe period in rhythm method, 68
Developed countries
annual rate of growth of, *10*
developing countries and,
crude birth rate of, *11*
crude death rate of, *11*
current fertility patterns of, *12*
females of, demographic characteristics of, *12*
per capita energy consumption of, *16*
per capita gross national product of, *16*
projected population increase of, *11*
growth of, problem of, 11
total population of, *10*
Developing countries
developed and. *See* Developed countries.
family planning programs in new contraceptive users in, *106*
growth of, problem of, 11
Diabetes, oral contraceptives and, 91
Diagnosis of birth defects with legalized abortion, 133
Diaphragm
as a contraceptive, 74–77
insertion of, 75, 76
manually, *75*
positions for, *74*
with introducer, *76*
removal of, *76*
spermicidal preparation used with, 75
Differential fertility, definition of, 1
Dilatation and curettage as abortion technique, 144
Dilatation and evacuation suction as abortion technique, 144
Dilation, cervical, with laminaria tents, 145
Discontinuation of oral contraceptives, reasons for, 96
Domestic alternatives, lack of, as barrier to woman's equality, 40
Douche, postcoital, 70
Down's syndrome, age of mother and, *36*
Drugs, development of embryonic system and, *91*
Dysmenorrhea
intrauterine device and, 104
oral contraceptives and, *83*, 94

ECONOMIC factors in large family, 31
Economic growth, declining fertility and, *18*
Economic problems, social and, of world population growth, 15–18
Education
as changing barrier to woman's equality, 41
as fertility factor, 2
in United States, 27
contraception and, *50*
sex, of adolescents, 53
unwanted fertility and, *60*
Effectiveness of contraceptive utilization, 60–67
Elements of demography, 3–5
Embryonic systems, development of, *91*
Emigration as a solution for limiting world population growth, 19
Emotional factors in female sterilization, 114
Endometrial aspiration as abortion technique, 146
Endometriosis, oral contraceptives and, 94
Endometrium
activity of, during menstrual activity, *67*
cancer of, mortality from, *90*
changes in, in various cycles, *80*
effects of the pill on, 81
Energy consumption, per capita, *16*
Epididymal spermatazoa maturation, retardation of, as means of contraception in future, 117
Equality of woman. *See* Woman's equality.
Ethnicity as fertility factor, 2
Evaluation of data, collection and, as an element of demography, 3, 4
Expulsion of intrauterine device, 103
Extent of adolescent pregnancy, 54–55

Extra-amniotic prostaglandins, uterine, as midtrimester abortion technique, 149

FACILITIES for adolescent pregnancy, 57
Factor(s) affecting fertility
age as, 2
demographic, 1–6
education as, 2
ethnicity as, 2
in United States, 26–29
income as, 3
marital patterns as, 2
migration as, 3
occupation as, 3
residence as, 2
Failed abortion, 142
Failure of contraceptive methods, 64–67
Family
concept of, in woman's changing status, 40
health consequences on, size and, 32–34
large, economic factors in, 31
population stresses on, 31–37
size of
economic factors in, 31
health consequences of, 32–34
3-child vs. 2-child, population growth with, *25*
Family income
food and expenditure and, *32*
medical expenditures and, *32*
Family planning
associations for, recommendations for, 95
KAP-gap in, 66
services of, as a solution to the world population problem, 20
statement concerning, by International Planned Parenthood Federation, 95
Family Planning Centers as solution for high risk patients, 36
Family planning programs
countries with, growth of, *20*
in developing countries, new contraceptive users in, *106*
Family size
food expenditure and, *32*
intelligence test scores and, *33*
medical expenditures and, *32*
Females
adolescent
live births to, *54*
never-married, contraceptives and, *50*
sexually active, *47*
coital frequency of, *49*
demographic characteristics of, *12*
ever-married, source of contraception and, *52*
never-married, source of contraception and, *53*
Female sterilization, 108–114. *See also* Sterilization.
sequelae to, 114
Fertility
declining, economic growth and, *18*
demographic factors in, 1–6
factors affecting, in United States, 26–29
trends in, in United States, 24
unwanted, education and, *60*
Fertility behavior changes, long-range approach to, *19*
Fertility control. *See also* Contraception; Contraceptive methods.
birth rates and, *62*
Fertility patterns
as world population problem, 12
current, *12*
Fertility rates
poverty status and, *26*
race and, *26*
Fetal death. *See* Infant mortality.
Fetal death rates, birth orders and, *35*
Fimbriectomy, 109
First pregnancy, complications of, *35*
First trimester abortion
complications of, 142
techniques for, 144–146
Foams as contraceptive method, 71–73
Food expenditures by family size and income, *32*
Food production, percent increase of, *17*
Future concerns in world population, 9–18
Future reproduction, abortion and, 136

GENERAL fertility rate, definition of, 5
Genitourinary system, oral contraceptives and, 81–83
Geographic distribution of world population, *13*
Gonadotropins, effect of oral contraceptives on, 81
Government and politics in woman's changing role, 40
Gross national product, per capita, *16*
Gross reproduction rate, definition of, 5
Growth of urban population, 18

HEALTH
abortion and, 136
affected by family size, 32–34
affected by short birth intervals, 34
child, affected by family size, 32–33
maternal
at extremes of reproductive years, 34–36
short birth intervals and, 34
parental, affected by family size, 33
Health care provider in abortion services, 151
Hemorrhage with intrauterine saline instillation, 148
High risk patients. *See also* Reproductive years, extremes of.
oral contraceptives and, 86, 89, 90
solutions for, family planning centers as, 36
Hormones, changes in, with various cycles, *80*
Hypernatremia with intrauterine saline instillation, 148
Hyperpigmentation, oral contraceptives and, 81
Hypertension
mechanism of, 89
oral contraceptives and, 89
Hysterectomy
as midtrimester abortion technique, 150
for sterilization, 114
Hysteroscopy for sterilization, 114
Hysterotomy as midtrimester abortion technique, 150

IDIOPATHIC pulmonary embolism, death from, *88*
Illegal abortions in various countries, *124*
Illegitimacy
as fertility factor in United States, 27
decrease in, with legalized abortion, 131
Illegitimate births to teenagers, *54*
Illiteracy rates, adult, *2*
Immigration as fertility factor in United States, 27, 28
Immunologic control as means of contraception in future, 117
Implants, polymer, as a contraceptive, 100
Implications of population problems in United States, 24
Imports, dependency on, by U. S., *23*
Income as fertility factor, 3
Incomplete abortions, decline in, with legalized abortion, 130
Infant. *See* Children.
Infant mortality, *131*
birth order and social class in, *33*
birth order in, *36*

Infant mortality—*Continued*
decline in, with legalized abortion, 131
social class and birth order in, *33*
Infection, intrauterine device and, 104
Infertility, Depo-Provera and, 99
Insertion
of diaphragm, 75, 76
of intrauterine device, 105
Institutional definitions as barrier to woman's equality, 40
Intelligence quotient
family size and, *33*
parental age and, *56*
Intelligence test scores, family size and, *33*
Internal migration as factor affecting fertility in United States, 28–29
International Planned Parenthood Federation, Central Medical Committee of, statement by, 95
Intra-amniotic prostaglandins as midtrimester abortion technique, 149
Intramuscular injection as contraceptive method, 98
Intramuscular prostaglandins as midtrimester abortion technique, 150
Intrauterine device
anemia and, 104
as contraceptive, 102–106
bleeding and, 104
continuation rates of, 102
contraindications to, 104
copper, 105
discontinuation of, reasons for, *103*
dysmenorrhea and, 104
expulsion of, 103
infection and, 104
insertion and retention of, 105
pelvic pain and, 104
pregnancy and, 104
retention of, 105
samples of, *101*
side effects of, 102
usage of, *64*
uterine perforation with, 103
Intrauterine extra-amniotic prostaglandins as midtrimester abortion technique, 149
Intrauterine progesterone contraceptive system, 106
Intrauterine saline instillation
as abortion technique, 147–148
complications in, 147–148
mechanisms of action of, 147
physiologic changes during, 147
retained placenta with, 148
substitutes for, 148
Intravaginal prostaglandins as midtrimester abortion technique, 149
Intravenous prostaglandins as midtrimester abortion technique, 149
IPPF, statement by, 95
IUD. *See* Intrauterine device.

JELLIES as contraceptive method, 71–73

KAP-gap, *66*
in contraceptive failure, 66
Kingsley Davis theory, 4
Knowledge
of conception by unmarried young females, *53*
of contraception, in failure of, 66
of reproduction and conception by adolescents, 53

LABORATORY alterations, oral contraceptives and, 84
Lactation
Depo-Provera and, 99
effect of, on normal function, *71*
oral contraceptives and, 97
prolongation of, as a contraceptive method, 70
Laminaria tent
as cervical dilator, 145
hygroscopic swelling of, *145*
placement of, *145*
Laparoscopy sterilization, *112*
Laparoscopy tubal ligation, 113
Legal institutions as barrier to woman's equality, 40
Legalized abortion
advantages of, 128–133
by age, *143*
child abandonment and, decline of, 131
diagnosis of birth defects in, 133
disadvantages of, potential, 133–136
in U. S., *128*
increase in, potential for, 133–136
mortality in, by method, *138*
population pressures and, easing of, 128
psychological problems in, 135
survey of, *127*
women seeking, profile of, *144*
Legislation, abortion, 123–127
in the United States, 124–127
Libido alterations with oral contraceptives, 81
Life, quality of, effects of contraception on, 63–64
Lippes loop in utero, *102*
Live births
number of, change in, *129*
to adolescent females, *54*
Liver, oral contraceptives and, 90
Low-birth-weight infants, age of mother and, *36, 56*
LRH in rhythm method, 69
Luteinizing hormone releasing factor in rhythm method, 69
Luteolytic agent(s)
as means of contraception in future, 119
map of, 118

MALE sterilization, 114–117
complications in, 114
Malthusian theory, 4
Management, office, of oral contraceptives, 94
Marital patterns as factor in fertility, 2
Marital status, induced abortions and, *143*
Marriage
as barrier to woman's equality, 40
as fertility factor in United States, 26
in United States, 26
Marxist theory, 4
Mastalgia, oral contraceptives and, 81
Maternal consequences of short birth intervals, 34
Maternal death rates, *35*
Maternal depletion syndrome, short birth intervals and, 34
Maternal mortality, decline in, with abortion, 129–130
Mechanisms of action
in oral contraception, 79
of copper IUD, 105
of Depo-Provera, 98
of intrauterine saline instillation, 147
of morning-after-pill, 100
of polymer implants, 100
Medical care, expenditures for, by family size and income, *32*
Medical consequences of adolescent pregnancy, 55
Medical facilities
as asset to woman's equality, 42–43
as barrier to woman's equality, 42–43
Medical knowledge
as asset to woman's equality, 42–43
as barrier to woman's equality, 42–43
Medicolegal documentation in rape, 43
Menarche, declining age of, *55*
Menorrhagia, oral contraceptives and, 94

Menstrual cycle
length of, *68*
ovarian and endometrial activity during, *67*
Menstrual extraction
as abortion technique, 146
equipment for, *146*
Menstrual regulation as abortion technique, 146
Menstruation
return of
after abortion, *134*
postpartum, *71*
Methodological tools as an element of demography, 3, 4
Methods
of abortion. *See* Abortion.
of contraception, 59–122. *See also* Contraceptive methods.
Metreurynter as midtrimester abortion technique, 150
Midtrimester abortion, technique for, 147–150
Migration
as fertility factor, 3
internal, as fertility factor in United States, 28–29
Mini-abortion as abortion technique, 146
Minilaparotomy, *113*
suprapubic, 114
Minipills, 98
Minorities as fertility factor in United States, 26–27
Morbidity
febrile, with intrauterine saline instillation, 147
in therapeutic abortions, 139–141
Morning-after-pill, 100
Mortality
abortion and, *129, 130*
by gestation and procedures, *138*
by method, *138*
by weeks of gestation, *137*
as world population problem, 9–10
cancer and, *90*
decline in
infant, with legalized abortion, 131
maternal, with legalized abortion, 129-130
in children, order of pregnancy and, *32*
in intrauterine saline instillation, 147
in therapeutic abortions, 136–139
infant
by age of mother, *56*
decline in, with legalized abortion, 131
short birth intervals and, *34*
maternal
and abortion mortality rates, *130*
decline in, with abortion, 129–130
puerperal, abortions and, *129*
Myocardial infarction with oral contraceptives, 90
Myometrial necrosis with intrauterine saline instillation, 148

NET reproduction rate, definition of, 5
Never-married adolescent females, contraceptives and, *50*
Nonseptic abortion rates, *130*
Nulligravida, intrauterine device for, 105
Nulliparas, copper IUD and, 105

OCCUPATION
as fertility factor, 3
of husband, children ever born to, *3*
Occupational distribution with sex-typing of jobs, *42*
Occupational experiences
and woman's changing role, 41
effect of fertility on, 41
Office management of oral contraceptives, 94
Oral contraception, 77–97
coital frequency and, *77*
omission of, pregnancy rates in, *96*
Oral contraceptives
and changes in endometrium, ovary, and plasma hormone, *80*
anemia and, 94
blood pressure and, rise in, 89
breast tenderness and, *83*
cancer and, 90
cerebrovascular accidents and, 88
complaints and, *83*
diabetes and, 91
discontinuation of, reasons for, 96
dysmenorrhea and, *83*
effects of
on breasts, 81
on cardiovascular system, 83
on central nervous system, 81
on cervix, 81
on endometrium, 81
on genitourinary system, 81–83
on gonadotropins, 81
on liver, 90–91
on ovary, 81
on reproductive system, 80–81
on skin, 81
on vagina, 80
physiologic, 80–81
hypertension and, 89
mechanisms of action in, 79
myocardial infarction and, 90
non-medical distribution of, 94
office management of, 94
physiologic effects of, 80–81
postpill amenorrhea and, 92
pregnancy and, 92
premenstrual tension and, 94
prescriptive techniques with, 95
risks involved with, 86
side effects of, 81–84, *84*
skeletal maturation and, 93
steroids in, *78*
summary of, 93
teratogenicity and, 91
thromboembolic disease and, 86–89
usage of, *64*
vitamin deficiency and, 93
Oral prostaglandins as midtrimester abortion technique, 150
Out-of-wedlock births. *See* Illegitimacy.
Ovarian activity during menstrual cycle, *67*
Ovary
changes in, in various cycles, *80*
effects of oral contraceptives on, 81
Ovulation
return of
after abortion, *134*
postpartum, *71*
Ovulatory cycle, length of, *68*

PAIN, pelvic, intrauterine device and, 104
Parental health affected by family size, 33
Parity, sterilization and, 108
Patient
characteristics of, in abortion, 143–144
in abortion, 150, 151
Patient selection, oral contraceptives and, 97
Pearl formula, 65
Pelvic pain, intrauterine device and, 104
Period fertility, definition of, 5
Perforation, uterine, intrauterine device and, 103
Physical factors in female sterilization, 114
Physical injuries, care of, in rape, 43
Physiological changes during intrauterine saline instillation, 147
Physiological effects of the pill on the reproductive system, 80–81
Pill, 77–97. *See also* Oral contraception; Oral contraceptives.
Pill-a-month, 99
Placenta, retained, with intrauterine saline instillation, 148

Politics, government and, in woman's changing role, 40
Polymer implants, 100
Pomeroy tubal ligation, 109
Population
  annual percentage increase of, *9*
  as an ecosystem variable, 3
  of world. *See* World population.
  total, *10*
  United States
    anticipated, by age groups, *25*
    future, with 3-child or 2-child family, *25*
    problems and implications of, 24–25
  geographic distribution of, 13
Population dynamics, 1–30
Population growth, 9, 10
  annual rate of, *10*
  in United States sunbelt region, *28*
  projected percent increase of, *11*
  through history, *9*
  world, limiting of, solutions for, 19
Population growth rate
  doubling time and, *9*
  years to double and, *5*
Population pressures, easing of, with legalized abortion, 128
Population problems
  in the United States, 23–29
  in the world, 7–22
Population pyramid, *16*
Population stresses on the family unit, 31–37
Postabortal period, intrauterine device and, 104
Postcoital contraception, morning-after-pill as, 100
Postcoital douche, 70
Postovulatory contraception, morning-after-pill as, 100
Postpartum period, intrauterine device and, 104
Postpartum return of menstruation and ovulation, *71*
Postpartum program, pregnancy rates and, *61*
Postpill amenorrhea, 92
Poverty
  as fertility factor in United States, 27
  fertility rates and, *26*
Precautions in prescribing oral contraceptives, 96
Practice of family planning in failure of fertility control, 66
Pregnancy(ies)
  among adolescents. *See* Adolescent pregnancy.
  first, complications of, *35*
  intrauterine device and, 104
  oral contraceptives and, 92
  order of, childhood mortality rates and, *32*
  prior, induced abortions and, *143*
  symptoms of, oral contraceptives and, 83
Pregnancy interception, morning-after-pill as, 100
Pregnancy prevention. *See also* Contraceptive methods.
  in rape, 44
Pregnancy rates
  oral contraceptive omission and, *96*
  postpartum program and, *61*
Pregnancy related complications in extremes of reproductive years, 35
Pregnancy termination. *See* Abortion.
Premenstrual tension, oral contraceptives and, 94
Prescriptive techniques, oral contraceptives and, 95
Primary health care, access to, with legalized abortion, 132
Procreation as barrier to woman's equality, 40
Progesterone contraceptive, intrauterine, 106
Progestasert, 106
Prolongation of lactation as a contraceptive method, 70
Prostaglandins, *148*
  abortion rate and, *149*
  abortions and, *129*
  as midtrimester abortion technique, 148–150
  extra-amniotic, as midtrimester abortion technique, 149
  intra-amniotic, as midtrimester abortion technique, 149
  intramuscular, as midtrimester abortion technique, 150
  intrauterine extra-amniotic, as midtrimester abortion technique, 149
  intravaginal, as midtrimester abortion technique, 149
  intravenous, as midtrimester abortion technique, 149
  oral, as midtrimester abortion technique, 150
Psychological damage, prevention of, in rape, 44
Psychological problems
  in female sterilization, 114
  in legalized abortion, 135
  in vasectomy, 117

QUALITY of life, effects of contraception on, 63–64
Quinestrol as oral contraceptive, 99

RACE
  contraception and, *50*
  contraceptive methods and, *51*
  fertility rates and, *26*
  induced abortions and, *143*
  of husband, children ever born to, *3*
  of mother, children ever born to, *1*
  sterilization and, *107*
Racial differences in social patterns of adolescents, 49
Rape
  care of physical injuries in, 43
  care of victim in, 43
  health care worker in, 43–44
  medicolegal documentation in, 43
  morning-after-pill and, 102
  pregnancy prevention in, 44
  psychological damage in, prevention of, 44
  venereal disease prophylaxis in, 44
  Women Organized Against, 43, 44
Rate of population increase, definition of, 4
Reanastomosis
  in vasectomy, 117
  tubal, following laparoscopy tubal ligation, 113
Recanalization
  in laparoscopy tubal ligation, 113
  in vasectomy, 115
Relocation as solution for limiting world population growth, 19
Replacement fertility, 25
  population projections based on, *10*
Replacement level, transition from uncontrolled fertility to, *129*
Reproduction
  future, abortion and, 136
  knowledge of, by adolescents, 53
Reproductive system, physiological effects of oral contraceptives on, 80–81
Reproductive years
  extremes of, in fertility rates, 34–36
  extremes of, pregnancy related complications in, 34–36
Residence as fertility factor, 2
Retained placenta with intrauterine saline instillation, 148
Retardation of epididymal spermatazoa maturation as means of contraception in future, 117
Retention of intrauterine device, 105
Rhythm method, 68
  basal body temperature in, 68
  cervical mucus in, examination of, 69
  luteinizing hormone releasing factor in, 69
Rights, human, declaration of, 7

Risks involved with oral contraceptives, 86
Rivanol as midtrimester abortion technique, 150
Rubber as contraceptive method, 73

SAFE period, oral contraceptives and, 97
Saline abortion
  abortion rate and, *149*
  complications of, in second trimester, 142
Second trimester saline abortion, complications of, 142
Septic abortion rate, *130*
Septic abortions, decline in, with legalized abortion, 130
Sex-typing of jobs, *42*
Sexual assault. *See* Rape.
Sexual patterns of adolescents, 47–53
Sexuality
  adolescent, 47–57
  female, 39–45
  male, after vasectomy, 117
Sexually-active adolescent females, *47*
Sexually-active adolescents, contraceptive methods and, *52*
Sexually-experienced never-married females, contraceptive methods and, *52*
Short birth intervals
  health consequences in, 34
  mortality and, *34*
Side effects
  adverse, of oral contraceptives, *82*
  beneficial, of oral contraceptives, *82*
  of Depo-Provera, 98
  of intrauterine device, 103
  of morning-after-pill, 100
  of oral contraceptives, 81–84, 93
  of the pill, 81–84, 93
  oral contraceptives and, *84*
Skeletal maturation, adolescents and, with oral contraceptives, 93
Skin, effects of oral contraceptives on, 81
Social class, infant mortality and, *33*
Social consequences of adolescent pregnancy, 55
Social customs as barriers to woman's equality, 40
Social patterns
  of adolescents, 47–53
    racial differences in, 49
Social problems
  economic and, of world population growth, 15–18
  encountered by working women, 41
Social Security Amendment of 1972, family planning services under, 36
Socioeconomic factors, legalized abortion and, 132
Solutions
  for high risk women, family planning centers as, 36
  for limiting population growth of the world, 19
Spermicidal action of copper IUD, 105
Spermicidal chemicals in contraception, 71
Spermicidal preparation(s)
  application and distribution of, *73*
  used with the diaphragm, 75
Sterilization
  age factor in, 107
  age range in, *107*
  as contraceptive method, 106–117
  by laparoscopy, *112*
  by wife's age, *107*
  chemical occlusive agents in, 114
  female, 108–114
  incidence of, 106
  male, 114–117
  parity and, 108
  type of, in United States, race and, *107*
Steroidal suppression of spermatozoa as means of contraception in future, 117
Steroids in oral contraceptives, *78*
Sunbelt region, population growth in, *28*
Suppositories as contraceptive method, 71–73
Suprapubic minilaparotomy, 114
Surgical contraception. *See* Sterilization.
Symptoms of pregnancy, oral contraceptives and, 83

TECHNIQUES of abortion, 144–150
Teenagers. *See* Adolescent(s).
Teratogenicity
  morning-after-pill and, 102
  oral contraceptives and, 91
Termination of pregnancy. *See* Abortion.
Theoretical issues as element of demography, 3–4
Theory(ies)
  in demography, 3–4
  Kingsley Davis, 4
  Malthusian, 4
  Marxist, 4
Therapeutic abortions
  complications of, 136–143
  morbidity in, 139–141
  mortality in, 136–139
  risks and complications of, 136–143
Therapeutic benefits from oral contraceptives, 94
Thromboembolic disease
  etiology of, 87
  evidence of, with oral contraceptives, 88
  incidence of, with oral contraceptives, 88
  oral contraceptives and, 86–89
    statement concerning, 95
Thrombophlebitis. *See* Thromboembolic disease.
Total fertility rate, definition of, 5
Total population, *10*
Transition, demographic, 5
  stages of, *5*
Trends in fertility in United States population problems, 24
Tubal ligation
  abdominal, 108
  laparoscopy, 113
  Pomeroy, 109
  types of, and results, *110*
  vaginal, 109
Tubal reanastomosis following laparoscopy tubal ligation, 113

UN DECLARATION of Human Rights, 7
Uncontrolled fertility, transition from, to replacement level, *129*
United States
  births in, to mothers under 18 and over 35, *35*
  fertility in
    education and, 27
    factors affecting, 26–29
    minorities and, 26
    poverty and, 27
    trends in, 24
  illegitimacy and, 27
  immigration in, 28
  imports by, *23*
  internal migration in, 28–29
  marriage in, 26
  population of. *See* Population, United States.
  population problems in, 23–29
  sunbelt region of, population growth in, *28*
Urban population, growth of, *18*
Uterine perforation, IUD and, 103
Uterine rupture with intrauterine saline instillation, 148
Utilization of contraceptives. *See* Contraceptive utilization.

VACTERL ANOMALIES, oral contraceptives and, 91
Vagina, effects of the pill on, 80
Vaginal tubal ligation, 109

Vasectomy
complications in, 115–117, *116*
for fertility control, 114–117
recanalization in, 115
reversibility and, 115–117
psychological problems in, 117
technique of, *115*
Venereal disease prophylaxis in rape, 44
Vitamin deficiency, oral contraceptives and, 93
Vocabulary for demographic studies, 4

WATER intoxication with intrauterine saline instillation, 148
Welfare payments, fertility rate and, 27
Withdrawal as contraceptive method, 70
WOAR. *See* Women Organized Against Rape.
Woman, changing status of, 39–45
Woman's equality
and how it affects fertility, 39–45
changing barriers to, 39–43
church as barrier to, 40
cultural labels as barrier to, 40
changing status of, 39–45
educational opportunities and, 41
family and, 40
government and politics as barrier to, 40
institutional definitions as barrier to, 40
lack of domestic alternatives in, 40
legal institutions as barrier to, 40
marriage and, 40
medical facilities and, 42
medical knowledge and, 42
occupational experiences and, 41–42
politics and government as barrier to, 40
procreation as barrier to, 40
social customs as barriers to, 40
Woman's role
as barrier to woman's equality, 40
customs in, 39
Women
adolescent. *See* Females, adolescent.
as nonfarm work force, *42*
married, who work, percentage of, *42*
seeking abortion, profile of, *144*
Women Organized Against Rape
services of, 44
statistics from, 43
Women's liberation and how it affects fertility, 39–45
Working women, social problems encountered by, 41
World population
developed and developing countries and. *See* Developed countries, developing countries and.
economic and social results of, 15–18
family planning services and, 20
fertility patterns in, 12
future concerns in, 9–18
geographic distribution of, *13*
growth of, limiting of, solutions for, 19
high dependency ratios in, 15
limitation of, solutions for, 19–20
mortality and, 10–11
problems of, 7–22
factors affecting, 9–18
projections of, *10*
social and economic results of, 15–18
urban, growth of, *18*
World Population Conference, 1974, final statement from, 7–8

YEARS to double population
definition of, 5
growth rate and, *5, 9*

ZERO population growth, definition of, 5
ZPG, definition of, 5